Spaced Out

A Casual Approach to the Cosmos

By
Kalpana Pot

ISBN: 9798870921556

Printed and bound in the United States of America by Amazon.com

Find Kalpana on social media.
@KalpanaPot

Table of Contents

Introduction Pg 05
Prologue Pg 06
Preface Pg 09
Pale Blue Dot Pg 10
The Unknown Pg 11
- Wormholes Pg 12
- White Holes Pg 13
- Multiverse Pg 14
- String Theory Pg 15
- Dark Photons Pg 16
- Kardashev Scale Pg 17
- Fermi Paradox Pg 18
- The Great Filter Pg 19
- Time Travel Pg 20
- Simulation Pg 21
- Panspermia Pg 22
- Women in STEM Pg 23
- Women in STEM 2 Pg 24

Solar System Pg 25
- Solar System Basics Pg 26
- Solar System Structure Pg 27
- Discoveries Pg 28
- Formation Pg 29
- The Sun Pg 30
- The Hottest Star Pg 31
- Solar Activity Pg 32
- Rocky Planets Pg 33
- Gas Giants Pg 34
- Kepler's Laws Pg 35
- Uranus's Tilt Pg 36
- Spacecraft Pg 37
- International Space Station Pg 38
- Moons Pg 39
- Our Moon Pg 40
- To the Moon Pg 41
- The Fallen Pg 42
- Eclipses Pg 43
- Zodiac Pg 44
- Asteroids, Comets, Meteors, oh my! Pg 45
- Armageddon Pg 46
- Recipe for Life Pg 47
- Life Pg 48
- Edge of the Solar System Pg 49
- Pluto Pg 50
- Planet IX Pg 51

Stars Pg 52
- A Star is Born Pg 53
- Stellar Classification Pg 54
- HR Diagram Pg 55
- A Star is Dead Pg 56
- Supernovae Pg 57
- Weird Stars Pg 58
- Stardust Pg 59

Galaxies Pg 60
- The Milky Way Pg 61
- Hubble's Law Pg 62
- Galaxy Types Pg 63
- Measuring Distances Pg 64

Exoplanets Pg 65
- Exoplanet Overview Pg 66
- Finding Exoplanets Pg 67
- Weird Exoplanets Pg 68

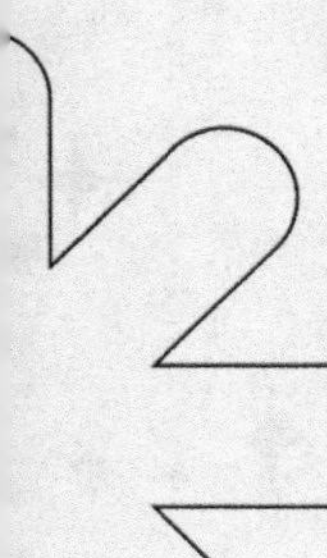

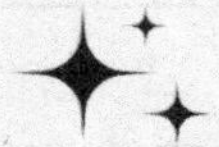

Table of Contents

Telescopes Pg 69
- Electromagnetic Spectrum Pg 70
- Spectroscopy Pg 71
- Light, Fast AF Pg 72
- Dark Skies Pg 73
- Types of Teles Pg 74
- The OGs Pg 75
- Next Gen Pg 76
- Ground Teles Pg 77
- Spacetime Pg 78
- Cosmological Constant Pg 79
- The Big Bang Pg 80
- Big Bang Evidence Pg 81
- Redshift Pg 82
- Dark Matter & Dark Energy Pg 83
- Observable Universe Pg 84
- End Theories Pg 85
- Geometry of the Universe Pg 86
- Thermodynamics Pg 87
- The End Pg 88

Interstellar Pg 89
- Nebulae Pg 90
- WTF is a Black Hole? Pg 91
- Black Hole Structure Pg 92
- Black Hole Types Pg 93
- Black Hole Breakthroughs Pg 94
- Quasars Pg 95
- Gravitational Waves Pg 96
- Gamma-Ray Bursts Pg 97
- Fast Radio Bursts Pg 98
- LFBOTs Pg 99
- JuMBOs Pg 100

Relativity Pg 101
- Law of Universal Gravitation Pg 102
- Relativity Pg 103
- Special Relativity Pg 104
- Time Dilation Pg 105
- General Relativity Pg 106
- Gravitational Lensing Pg 107
- Newton vs. Einstein Pg 108

Quantum Pg 109
- The Atom Pg 110
- Quantum (fml) Pg 111
- Quantum Mechanics History Pg 112
- Atom Bomb Pg 113
- Four Forces Pg 114
- Standard Model Pg 115
- Antimatter Pg 116
- Superposition Pg 117
- Double-Slit Experiment Pg 118
- To See or Not to See... Pg 119
- Entanglement Pg 120
- Theory of Errythang Pg 121

Debunking Pg 122
- Flat Earth Pg 123
- Moon Landing Pg 124
- CGI Pg 125
- Defunding Pg 126

Final Thoughts Pg 127
Quick Facts Pg 128
Quick Facts Pg 129

Introduction

ABOUT KALPANA

Kalpana Pot is an actor, writer, host, and science communicator. Many people know her from her social media pages, where she delivers astronomy topics in a fun and unfiltered way. Her passion for the subject fuels her breadth of knowledge. After moving to LA, she was hired as a guide at the famous Griffith Observatory. Her job was to enlighten the public on the wonders of the Universe. (That's her description, not theirs.) Then, when the pandemic hit, like many others, Kalpana hopped on social media to continue speaking about space. Her online presence garnered numerous hosting opportunities, including working with Franklin Institute, Amazon, History Channel, and more. This is Kalpana's first book and is a non-fiction overview of everything space-related. Ages 12+.

Disclaimer Unless otherwise stated, assume everything that's talked about is round.

Prologue

It was her first day as a tour guide in her city's local space museum. She was nervous but excited. All her life, she loved space. She loved the pretty pictures, the mystery, and the endless possibilities the vast cosmos represented.

Her childhood home life was tumultuous, but the Universe offered her a silent escape. She could leave behind the screams and float among the stars. She could narrowly escape the grasp of a violent black hole. She could wander by a world where the beings were in perpetual peace. She would sit on her rooftop and gaze at the few stars strewn across the sky; a slight smile always replaced the few tears that streamed down her cheeks. She was home.

When she entered the museum, she was greeted by the guide in charge and was told to shadow the more established museum guides to see how things worked. She went from floor to floor, listening to her colleagues speak while taking in the array of visitors. These were people from all over the world who were told by some online website that this space museum was a "must-see." Not only did it offer the best views in the entire city, but many famous films used its beauty as a backdrop. At least, that's what most of the tourists seemed to be interested in. Did *anyone* there care about the science? She was becoming disheartened. On several occasions, it seemed like people would walk past her and address her male colleagues instead, even though they were all dressed in the same dorky space-themed uniforms.

Maybe I should smile more to seem more approachable, she thought. Nope. That didn't work. She deflates.

She decides to leave her post and wander around for a bit. Maybe she was being too hasty with her negativity. She came across a family, mom, dad, and teenage daughter checking out the interactive Periodic Table of Elements exhibit. Well, at least the parents were. The teenager was more interested in her phone. As the mom pressed buttons, she excitedly told her daughter to come and look at what was happening.

"Mom, I don't care about this stupid stuff!" snaps the daughter. Now, this was the last straw. She didn't become a museum guide to sit by idly. She approached the family.

"Did you guys hit the HUMANS button yet?" she asks. The mom hits the HUMANS button. Twenty-five element boxes on the top rows light up. "These are all the elements that makeup people. If you hit the SUPERNOVA button, these elements are made inside supernovae- insanely immense explosions caused by a massive star's death." She turns to the teenage daughter.

"If you noticed, that means all of the elements that make up everything we know, the silicon in your phones, the silver in your Tiffany bracelet, super cute, by the way, and even the iron in your blood all came from stars. Everything we care about, including ourselves, is literally made of stardust. We wouldn't be here without this *stuff*."

The teenage daughter smiles sheepishly. "I guess that's kind of cool." The parents thank her, and the family heads away. As she watches them leave, several visitors approach her, eager to ask questions. She smiles.

A few hours later, and now in a better mood, she takes her break. As she digs into her prepackaged salad, other employees enter. Operating a massive museum like this requires various employees, from traffic control to ticket sellers, museum guides to security, and PhDs to maintenance; the employees are as diverse and eccentric as those who visit. She introduces herself but suddenly gets cut off by a call on the security guard's radio. Wait, did they hear that right? The guard asks the person to repeat.

"We have a situation on the front lawn." says the caller. "Some people have shown up to, uh, protest." Confused and curious, they leave their food behind and head outside.

Across the lawn, a raucous group of people holding signs chant, "NASA lies, the Earth is flat" through megaphones. The employees roll their eyes. It's just another day at the space museum. They all head back inside except for one. She stands there, dumbfounded. Is this for real? Her anger begins to well.

"Wake up, sheeple," yells one protester. "These people be lyin' to you! The Earth's flat and the Moon landin' was a hoax! Don't let them gov't cronies fool you. Fight the new world order!"

"Umm, excuse me?" The protester turns around to see her. "What in the HELL are you talking about?" she asks.

The protester notes her uniform "You heard me, crony." he says aggressively. "Why y'all perpetuatin' these lies? The Earth's flat, and we NEVER went to no Moon!" The rest of the protesters gather around.

"Well, the Earth isn't flat, but your head might be!" she rebuttals. "How do you explain gravity being uniform everywhere on Earth if it wasn't round?"

"*Gravity's* a lie. The flat plane moves upwards." the protester says.

"Huh?! That doesn't make any sense!" she exclaims. How did you all find your way here? I assume you used GPS? GPS satellites are mathematically designed to overcome gravity as they speed AROUND the Earth."

"Ya think we're dumb enough to use government tracking devices?" the protester scoffs. "These *satellites* are as fake as them Moon landin's. And, I don't believe in no math!"

She palms her face in frustration. "Ok, let me tell you something."

Back in the museum, the rest of the guides go about their jobs. "Have you seen the new girl?" asks the lead guide. They shake their heads. The lead guide's eyes go wide.

"...Furthermore, there are over 8,000 photos, videos, and audio recordings. The Lunar Reconnaissance Orbiter, which still orbits the moon, has taken pictures of the landing site. Even Russia admitted we got there!" she explains frustratingly to the protesters.

"You expect us to believe some *orbiter*?" the protester asks. His group laughs. Other guides have now shown up.

A more timid guide approaches her, "It's best just to let it go."

The protester laughs. "Yeah, angry little woman, listen to your little f** friend."

In a flash, she punches the protester. Everything goes slo-mo.

"Ooohhhhhhh sssshhh*****," the lead guide exclaims. Both sides charge. Slo-mo stop. Chaos. It's an all-out brawl between guides and protesters, jumping on each others' backs, whacking each other with signs, etc. Her eyes widened as she took in the scene, which seemed to go on for eternity. She feels cuffed, not knowing what to do. No, wait, those are literal cuffs! The LAPD has shown up to break up the brawl. A few other guides and protesters are being cuffed as well. As she begins to hyperventilate, she gazes up at the all-too-familiar sky, searching for an escape.

This, my dear readers, is based on my true story. Yes, I love science, but I also love storytelling. My purpose in life is to bridge the two. I'm giving you a blurb on something I've been working on for a few years, hoping that if I release it to the Universe, it'll find its way into existence.
(FYI, I'm a screenwriter, not a novelist. My script is MUCH better.)

"Science is a genre of storytelling. Stories of the real world."
~ Sean Carroll

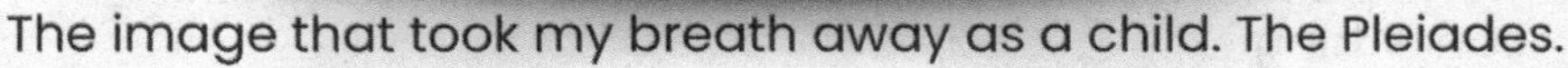

The image that took my breath away as a child. The Pleiades.

Preface

Astronomy is the branch of science that deals with celestial objects, space, and the physical Universe. For every subject I will speak about, there's a shit-ton of math to back it up. Obviously, I'm not delving into that unless absolutely necessary. I'm going over every topic I can think of that covers something to do with space, from the very large to the very small. I decided to write this book because I'm damn good at space science communication, and I want people to know they're never too old to learn anything that interests them. These topics are complex as hell, and I'm only scratching the surface to the best of my ability. Why do I want to? Because I love this shit and want everyone to feel the same way! As Plato said, "Astronomy compels the soul to look upwards and leads us from this world to another." I've always felt that studying the Universe is as much of a philosophy as a science. Knowing what is and isn't out there is a humbling experience. But with the increase in anti-science in recent years, which I can't wrap my head around, I feel more compelled than ever to advocate for science and the remarkable people in these fields doing remarkable things. These are the people who have gotten humanity to where we are today.

Thanks to science, life has become so comfortable that you can ironically shit on the science that made it that way. As Carl Sagan said, "We live in a society exquisitely dependent on science and technology, in which hardly anyone knows anything about science and technology." Thanks to studying the Universe, we have advancements in computers, cell phones, medical equipment, climate sciences, agriculture, and more! I'm not saying we shouldn't question anything. Questioning is the foundation of science. I'm saying put your brain power towards something more productive than arguing against the FACT that the Earth is round (as one example). I'm not sure where this mistrust and animosity towards "mainstream" science has come from, but it's become a calling for me to help mitigate this trend in a down-to-Earth way.

I also want to say thanks to my online audience! I have so much fun creating these space videos for you guys, and without you all, none of my sci-comm endeavors would exist. You've given me the confidence to write this book.

PS: before we begin, let me get this out of the way. The word *theory* gets thrown around too much in everyday life, but in science, a theory has evidence, has been tested, has been peer-reviewed, and hasn't been able to be disproven yet. Therefore, flat earth bullshit is NOT a theory; it's a (stupid) hypothesis, which is just a proposed explanation!

Pale Blue Dot

The Pale Blue Dot is a photograph of Earth taken Feb. 14, 1990, by NASA's Voyager 1 at a distance of 3.7 b miles (6 b km) from the Sun. It inspired Carl Sagan's book, "Pale Blue Dot: A Vision of the Human Future in Space". The following excerpt from his book is one of the most beautiful pieces ever written, in my opinion. When this world seems daunting, read this piece. If only every world leader had this *perspective*...

✶✦✶✦✶✦✶✦✶✦✶✦✶

Look again at that dot. That's here. That's home. That's us. On it everyone you love, everyone you know, everyone you ever heard of, every human being who ever was, lived out their lives. The aggregate of our joy and suffering, thousands of confident religions, ideologies, and economic doctrines, every hunter and forager, every hero and coward, every creator and destroyer of civilization, every king and peasant, every young couple in love, every mother and father, hopeful child, inventor and explorer, every teacher of morals, every corrupt politician, every "superstar," every "supreme leader," every saint and sinner in the history of our species lived there--on a mote of dust suspended in a sunbeam.

The Earth is a very small stage in a vast cosmic arena. Think of the rivers of blood spilled by all those generals and emperors so that, in glory and triumph, they could become the momentary masters of a fraction of a dot. Think of the endless cruelties visited by the inhabitants of one corner of this pixel on the scarcely distinguishable inhabitants of some other corner, how frequent their misunderstandings, how eager they are to kill one another, how fervent their hatreds.

Our posturings, our imagined self-importance, the delusion that we have some privileged position in the Universe, are challenged by this point of pale light. Our planet is a lonely speck in the great enveloping cosmic dark. In our obscurity, in all this vastness, there is no hint that help will come from elsewhere to save us from ourselves. The Earth is the only world known so far to harbor life. There is nowhere else, at least in the near future, to which our species could migrate. Visit, yes. Settle, not yet. Like it or not, for the moment the Earth is where we make our stand.

It has been said that astronomy is a humbling and character-building experience. There is perhaps no better demonstration of the folly of human conceits than this distant image of our tiny world. To me, it underscores our responsibility to deal more kindly with one another and to preserve and cherish the pale blue dot, the only home we've ever known.

— Carl Sagan, Pale Blue Dot, 1994

The Unknown

In screenwriting, the first ten pages are the most important. If you don't hook someone right away, they stop giving a damn and pass on your script. That's why I'm putting *The Unknown* section first... because this is the stuff about the Universe that people tend to love the most.

In the 1930s, Einstein and physicist Nathan Rosen used general relativity to propose the idea of "bridges" in space. They called them Einstein-Rosen bridges or <u>wormholes</u>.

Many sci-fi films and shows have referenced wormholes. The Universe is fucking huge, but hypothetical wormholes can act as a shortcut to get from one place in space to another.

The nomenclature is self-explanatory. Imagine a worm eating its way through an apple instead of going around it. Eating through it represents the shorter path. Since now you know that spacetime warps, wormholes could bend two parts of space towards each other. They could act as tunnels or make it as quickly as walking through a door and arriving at your destination.

However, wormholes might be unstable. If one wants to use one to traverse space, one must find a way to keep it open. This could be possible with <u>exotic matter</u>. Since gravity pulls regular matter together, perhaps there's a hypothetical exotic matter that does the opposite and repels. This could keep the wormhole from collapsing in on itself.

This is the stuff about the Universe that people love! The limits are only within our imagination, but I will reiterate that wormholes aren't real. Although general relativity mathematically predicts their existence, they remain purely hypothetical... for now.

There have been recent talks among scientists that wormholes and entanglement could be the same thing. Remember, two entangled particles can instantly affect each other at any distance. In a way, entanglement can be seen as a shortcut. So, if wormholes are a part of general relativity, entanglement is a part of quantum mechanics, and they *could* be describing the same thing... do you see where I'm going with this? Maybe this could be the *bridge* between the macro and the micro. In other words, maybe this could lead to a unifying theory.

White Holes

Since black holes, a region of spacetime where information can enter but never leave, exist - could their opposite, a region of spacetime where information exits but can never enter also exist? A white hole is mathematically possible!

The concept of white holes also came from general relativity and was developed by Austrian physicist Ludwig Flamm.

At this time, however, they are only hypothetical as no evidence has ever been observed anywhere.

There are some trippy thoughts behind the most exclusive club in the cosmos.

- They could be structured like a black hole: a singularity, an event horizon, a mass, maybe a spin, and material swirling around it, but nothing gets in.
- They repel matter that gets near.
- If you could reverse time while a black hole formed, you could get a white hole.
- No one knows how they could actually "form" since the reversal of a black hole is a massive star, and doing this would violate entropy (working against disorder).
- Our Universe has a time asymmetry, as in it only moves forward. Therefore, if white holes are backward black holes, they can never exist in our perception of the Universe.
- Objects inside could leave and interact with the outside, but nothing outside could interact with what's in it.
- Where does the mass inside come from if nothing can enter one?
- Could the Big Bang have been a white hole?
- Could entering a black hole lead you through a white hole in a different universe?
- Could a wormhole connect a black hole and a white hole?
- Could a white hole be at the very end of a black hole's life? Once it begins to die, it spits out everything it swallowed.
- Is a white hole a time machine where future information entering a black hole gets spat out in present time through a white hole?

As you can see, we get into sci-fi land regarding white holes, which makes them so fun! I'd like to imagine that if they did exist, our black holes would lead to these white holes in a different universe where time moves the other way.
And if the Big Bang was just a white hole, our Universe came from a black hole elsewhere.

Multiverse

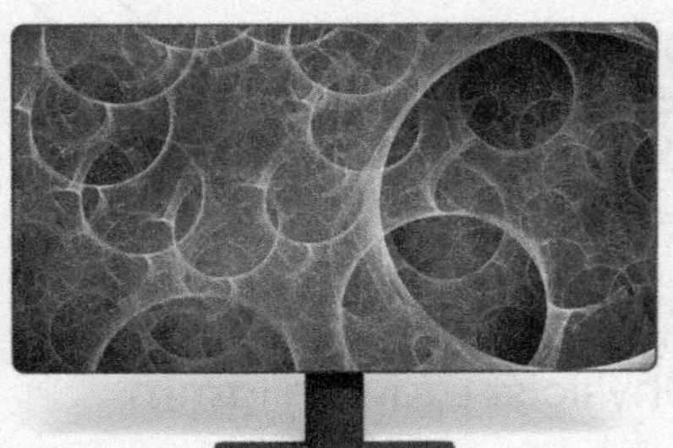

The idea behind the multiverse says that there are numerous (maybe infinite) universes outside of our own, and collectively they make up the multiverse. The concept makes for fun and creative storytelling, but as of now, it's just speculative. Each universe could have its own physics, chemistry, and biology laws. As usual, the scope of these universes is limited only by your imagination. But why did this concept arise?

- Inflation was the rapid expansion period of the Universe when it was less than a second old. Although inflation stopped soon after it began, it might not have stopped uniformly everywhere. If some parts kept going, they could also "pinch off" and become their own universes.
- Anthropic Principle: the fact that we have life in this Universe is like winning the lottery. So much had to line up correctly for life to arise here. Therefore, the odds only make sense if compared to a multiverse where, in most of the universes, it's not viable. But think about this... if there are an infinite number of universes, then there's the possibility of an infinite number of *yous*, even in a multiverse where life is rare.
- Everett's Many-Worlds Interpretation offers a different view of QM than the Copenhagen Interpretation. In superposition, the cat is alive or dead. In one case, you open the box, and thankfully, the cat is alive. The MWI says that while the cat is alive in your reality, a separate reality branched off simultaneously where the cat is dead. In this reality, I decided to write this book. The moment that decision was made, the reality where I decided *not* to write it branched off. Which reality has the more thankful audience, I can't say.
- The strange properties of dark matter and dark energy. Why the hell do we have an abundance of these enigmas that so greatly influence our Universe? We can't see them, and studying them is fucking hard. Perhaps these are properties of something from the outside... something from the multiverse.

Overall, no direct evidence of a multiverse has been found. There is no "damage" from our neighbor universe colliding with us and no indication that black holes are the connectors. It's a fascinating concept where you can draw your own philosophical conclusions.

String Theory

String theory tries to be an answer to the "theory of everything." I told you that unifying gravity with the quantum world will win you Nobel Prizes for a lifetime.

In the quantum section, we talked about the standard model of elementary particles: leptons, quarks, and bosons. These make up the particles that make up the atoms that make up matter. We kept thinking that we'd found the base of everything. Then string theory said, "Hold my drink."

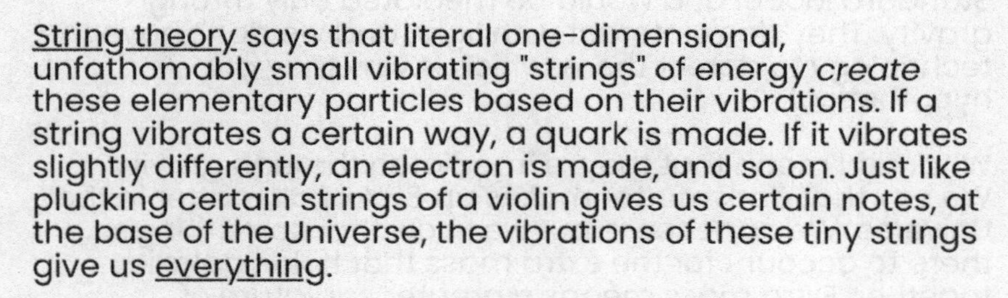

String theory says that literal one-dimensional, unfathomably small vibrating "strings" of energy *create* these elementary particles based on their vibrations. If a string vibrates a certain way, a quark is made. If it vibrates slightly differently, an electron is made, and so on. Just like plucking certain strings of a violin gives us certain notes, at the base of the Universe, the vibrations of these tiny strings give us everything.

Could strings also give us the long-awaited quantum gravity?

Introducing quantum gravity to how we currently understand elementary particles doesn't work. The math breaks down. However, If you do it through string theory, it does work! The math holds up when you introduce gravitons into the models.

But...

We experience the Universe in 4 dimensions. String theory only works mathematically in 10 dimensions (some variations say 11 dimensions). Although there is ample math behind this theory, ultimately, it's untestable, and that's a crucial part of substantiating anything in science.

Many continue to advocate for it, but it's lost some luster over time. But hey, it's doing more than I am when it comes to unification.

Dark Photons

"Dark photons" sounds like an oxymoron. Since photons are particles of light, what the hell are dark photons? They are hypothetical particles that aren't a part of the standard model but could explain dark matter, just like photons do for electromagnetism.

They would be a part of the hidden sector, a hypothetical collection of quantum fields and their corresponding particles that have yet to be observed! Particles in this sector would interact weakly with the particles in the Standard Model and would be mediated only through gravity. They don't interact with light, and we don't have the technology to detect them, which is why they are hypothetical.

What do I mean that they're "mediated through gravity?" We say that dark matter makes up 85% of all matter in the Universe. We can't see it, but we know that something's there to account for the extra mass that holds galaxies together. Extra mass means a greater curvature of spacetime, aka gravity. To sum it up, only gravity interacts with this dark stuff.

Only if/when hidden sector particles are discovered can they be added as an extension of the Standard Model. Then, they will also have a part to play in the "theory of everything."

Other hypothetical hidden sector candidates for dark matter particles are Weakly Interacting Massive Particles (WIMPs), axions, and sterile neutrinos, to name a few. (Please let *WIMPs* be a thing.) In this hidden sector, there could also be other Higgs bosons, other gauge bosons, gravitons that transcend different dimensions, exotic particles that explain dark energy, and particles that explain things we haven't even discovered yet. God, I fucking love space!

Kardashev Scale

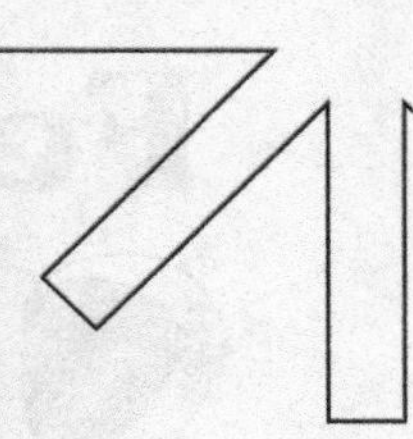

On Earth, humans consider themselves the most intelligent species. I suppose we've given ourselves that title because we've transformed our planet to fit our needs and comforts. We have a curiosity about our natural world that is greater than other species' levels of curiosity. One could argue, however, that the species living the most sustainably are the most intelligent. They aren't consumed by greed and aren't destroying their own home as a result. Apart from the fact that animals > people, my point is that advancement depends on the criteria.
In 1963, soviet astronomer Nikolai Kardashev came up with a scale to determine how advanced a civilization is, and it only depends on how much energy it uses. This is known as the Kardashev Scale.
There are generally 3 types of civilizations:
Type I civilizations harness all the energy from their home planets.

- Energy could come from fusion power, antimatter, solar energy, and any form of sustainable energy.
- They could manipulate the weather, volcanoes, and earthquakes, as all of the planet's power is in their control.
- Earth is considered to be a 0.72 on this scale.

Type 2 civilizations harness all the energy from their star.

- This is where the hypothetical Dyson Sphere comes into play. This structure would be built around the entire fucking star to collect its energy. Wait, it only gets crazier from here.

Type 3 civilizations can harness all the energy from their entire galaxy!

- At this point, a galaxy would be completely colonized, and energy would be extracted from all available sources.

This is where Kardashev himself stopped, as anything beyond is almost unimaginable. Almost.
Type 4 civilizations harness the energy from an entire Universe.

- I mean, at this point, they are just bored.
- The Universe's accelerating expansion may not mean anything to them, and they may even be able to live inside supermassive black holes.

Type 5 civilizations are basically gods, and the Universe is their playground. They can manipulate it as they please.
Type 6 civilizations can traverse the multiverse.
I'm going to stop here, but the limit, again, is only what's in your imagination. People have thought beyond these as it's fun and mindblowing!

Fermi Paradox

Developed by Enrico Fermi, the Fermi Paradox pretty much asks, "In such a vast Universe, where the hell is everyone?"

Take our Milky Way galaxy for example. We don't know its exact scale. There could be half a trillion or more stars, and on average, each star has at least one planet. Even if a small fraction of those planets are habitable and have life, that's still quite a high number! A small fraction in astronomical terms is pretty damn big. So again, in such a vast Universe, where the hell is everyone?

I already started to break down a part of this famous equation.

$$N = R_* \cdot f_P \cdot n_e \cdot f_l \cdot f_i \cdot f_c \cdot L$$

The Drake Equation, developed by astronomer Frank Drake in the 1960s, tries to ask the same question mathematically.

- *N*: the number of civilizations currently sending signals.
- *R**: the yearly formation rate of stars hospitable to planets.
- *fp*: fraction of those stars that have planets.
- *ne*: number of planets per solar system that have conditions suitable for life.
- *fl*: number of those planets where life actually develops.
- *fi*: number of those planets where life evolves to be intelligent.
- *fc*: number of intelligent life that develops technology that we could detect (radio signals).
- *L*: mean length of time that communicating life can communicate.

It sounds so simple. Just point dishes to space and wait for signals, right? No. Space is huge, and we don't scan the whole sky 24/7. My point is that there are numerous reasons why it's been radio silence, so to speak. Are we indeed alone, are we all in the same boat, or... are they avoiding us?

If it's the last option, one of my favorite reasons for avoidance is that we are the trailer trash planet. We could also be so primitive to them that they move right past us like we would an anthill. Once AGAIN, the reasons are only limited by your imagination. Personally, I think we're among the first.

The Great Filter

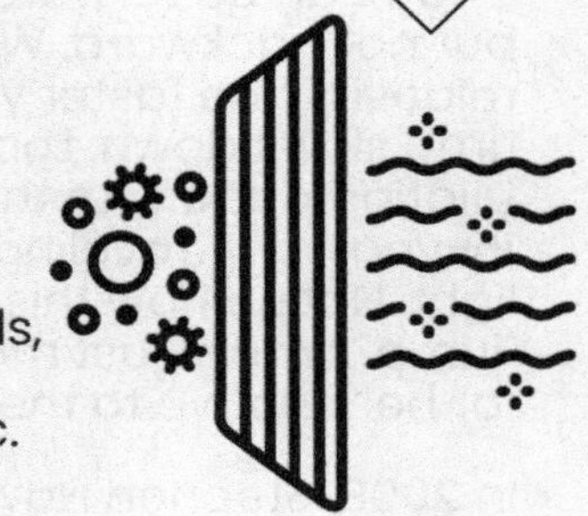

Okay, but what if we *are* alone? What would be the reason? The Great Filter is the idea that maybe there is some barrier that all life eventually hits as it makes its way to a Type 3 civilization.

Barriers could be from without: asteroids, gamma-ray bursts, aliens, etc. Or from within: climate change, war, viruses, etc. Basically, it's not easy for life to... life.

The filter is behind us.

The filter may be at the beginning. A planet could be habitable, but the filter generally occurs before life forms.

Life may form, but the filter occurs before jumping from prokaryotes to eukaryotes. To name a few differences, prokaryotes are single-celled and lack a nucleus. Eukaryotes can be single or multi-celled, have a nucleus, and are more complex than their predecessors. We're not talking about the biggest leap in evolution here, but it took almost two billion years to make that jump! So much crap could have happened in that time that would stop life from advancing further.

There have been 5 mass extinction events in Earth's history. The worst one is considered to be the Permian-Triassic extinction about 250 million years ago. Massive and widespread volcanic eruptions caused acidification of the oceans and extreme global warming that killed about 90% of all marine life and over 70% of all land life. Although it took a long time, life somehow made it through. But do you see how vulnerable life is?

The filter is ahead of us.

If we ever discover simple life elsewhere, then maybe the filter is *ahead* of us, and it's common for life to get to our stage... but not much further. If that's true, it would make the most sense that the cause of the filter would be due to that species' actions. Look at our current issues with climate change, AI, pandemics, and the constant threats of nuclear war. It would be sad to come this far and be the cause of our undoing, but that is the direction we seem to be heading in. Other civilizations may have already screwed themselves. Let's hope that won't be us.

Time Travel

You can go forward... to some extent, but not backward. With special relativity, the faster you go, the more time slows down. Look back at the Time Dilation page with the example of the 15-year-old traveling near the speed of light. Notice how this person didn't "jump" time. It just moved more slowly for her relative to the people on Earth.

In 2009, Stephen Hawking threw a party with a banner that read, "Welcome Time Travelers." Not one guest attended. His point was, if traveling back in time was possible, you bet your ass there would have been guests.
He had many thoughts on time travel because he understood the fundamental principles of our Universe. As in, the laws of the cosmos wouldn't allow for time travel. The "grandfather paradox" is a perfect example of that. If you traveled back in time and killed your grandfather, then ultimately, you wouldn't exist. If you didn't exist, then you didn't travel back in time and kill your grandfather. Going back in time and disturbing anything would alter something in the current time, so perhaps the laws of our Universe won't allow it.

We are governed by our three dimensions and the fourth dimension of time, which moves in one direction. Perhaps time could be transcended in hypothetical higher dimensions.

It would be amazing to go back and see history, or go forward and see how far science and technology have evolved past our lifetimes.

However, if there are no consequences for the decisions in your life because you can go back and change things... is there even a point to life? Do you ever learn? Do you ever grow? Is there a purpose if everyone's lives are now "perfect?" I hope that time travel means we can only observe and not alter.

Time travel is fascinating because it parallels the idea of immortality. Immortality speaks to our innate curiosity. We want to know how things turn out after we've left the party, and I totally get that. But, take comfort knowing you no longer exist once you've left the party. Therefore, your "longing" to know how things turn out wouldn't exist either. What I'm trying to say is, get over it.

Simulation

Are we in a simulation created by an unfathomably advanced civilization? Maybe.

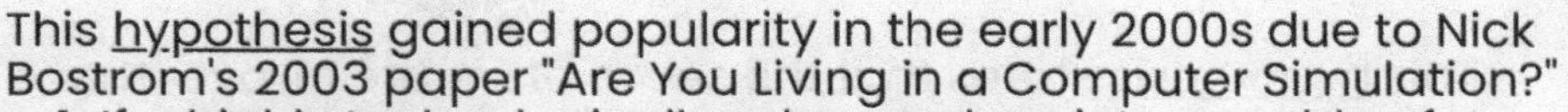

This hypothesis gained popularity in the early 2000s due to Nick Bostrom's 2003 paper "Are You Living in a Computer Simulation?"

1. If a highly technologically advanced society capable of creating powerful computers exists, they would likely have the capacity and motivation to simulate entire civilizations.
2. If this civilization could create such detailed simulations, they might create numerous simulations resembling their past or other hypothetical scenarios.
3. Given the potential for advanced civilizations to create simulations and the vast number of possible simulations, it's statistically more likely that we live in a simulation rather than the base reality.

Everyday "evidence" that people use to support it:

- The Mandela Effect: many people remember Nelson Mandela dying in the 80s even though he actually died in 2013. There are various instances in pop culture like this. Some say our "controller" is changing our past.
- Quantum mechanics: I mentioned the double-slit experiment where particles chose their state after they were observed. Maybe doing this saves processing power. And apparently, computer code has been found in equations of string theory.
- Laws of physics: some ask why our Universe has strict rules in the first place that are explained by math. Video games run the same way. The speed of light may be the processing limit of how information can be transmitted in our network.
- Paranormal activity: the creepy things we see are possibly glitches in the matrix.
- Human simulations: we are already pretty good at making simulations. From detailed maps of our Universe to the classic computer game The Sims (I killed many of them in the pool), we'd clearly be into simulating.
- Planck length is the smallest unit of length (10^-20 times smaller than a proton). This might be the pixel-sized building blocks of the Universe. It takes more Planck lengths to span a grain of sand than it would take grains of sand to span the observable Universe (mind blown).
- Clone stamp: there are repeating patterns in nature, perhaps to save on processing power.
- Ultimately, we can't prove we aren't in one!

Two personal points if we are in a simulation.

1. In this vast Universe, it *still* wouldn't be about us.
2. Who cares? How would it change your life? It wouldn't.

Panspermia

Pan: all. Sperma: seed.

This hypothesis states that the seeds of life are all over the Universe and are brought to planets by asteroids, dust, comets, and meteoroids. So then, life on Earth's true origins would have come from elsewhere.

There are variations to this hypothesis:

- **Directed Panspermia:** Life was deliberately spread to other planets by an advanced extraterrestrial civilization. The idea is that intelligent beings purposely sent microorganisms or even more advanced life forms to seed other worlds, potentially including Earth.
- **Lithopanspermia:** Life could have been transferred between planets within our own solar system via rocks ejected during impacts. Microorganisms trapped within these rocks could potentially survive the harsh conditions of space and eventually colonize new environments. (We try to avoid doing this when we send spacecraft to other bodies in our SS).
- **Radiopanspermia:** Life could be spread by radiation pressure from stars, such as through microbial spores being carried on dust grains.
- **Interstellar Panspermia:** Life could travel between star systems, either via natural processes like comet or asteroid impacts, or through more deliberate methods like probes or spacecraft.

"Life" could refer to microorganisms, advanced organisms, or more likely, the chemical building blocks needed for life to even exist. Of course, then you'd need the right environment for life to then thrive.

Some evidence for this hypothesis include:

- Extremophiles: microorganisms that are known to survive in very harsh conditions (including space).
- Organic molecules: we've found these and amino acids on comets, meteors, and in interstellar space.

The idea that life began elsewhere and was brought to Earth (deliberately or not) has been around for centuries. However, it remains purely a hypothesis leaving the origin of life on Earth still one of the biggest mysteries.

FYI: the prevailing hypothesis is abiogenesis - life arose from nonliving matter through natural processes, which technically doesn't rule out panspermia.

Women in STEM

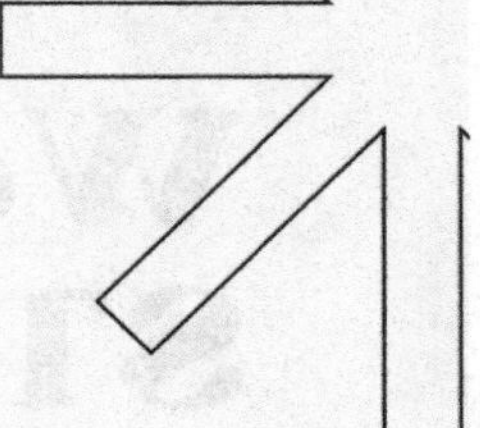

Sadly, it's quite fitting that I am putting the "Women in STEM" pages in the Unknown section. I'm not covering every topic in STEM of course, so here are only SOME of the remarkable women in space sciences throughout history.

- **Hypatia of Alexandria (c. 360–415)**:
  - An ancient Greek mathematician, astronomer, and philosopher who was considered the greatest mind of her time. She taught astronomy and mathematics at the Library of Alexandria. She suffered a violent death at the hands of Christians.
- **Caroline Herschel (1750–1848)**:
  - A German-born British astronomer who collaborated with her brother William Herschel. She discovered several comets and was the first woman recognized for her contributions to science when she was awarded the Gold Medal of the Royal Astronomical Society.
- **Maria Mitchell (1818–1889)**:
  - An American astronomer who became the first female professional astronomer in the United States. She discovered a comet in 1847 and was a pioneer in advocating for women's involvement in scientific research and education.
- **Henrietta Swan Leavitt (1868–1921)**:
  - An American astronomer who made groundbreaking discoveries related to Cepheid variable stars, which provided a crucial tool for measuring distances in the Universe and contributed to our understanding of the Milky Way's size and the scale of the cosmos.
- **Cecilia Payne-Gaposchkin (1900–1979)**:
  - A British-American astronomer who was the first to suggest that stars are primarily composed of hydrogen and helium, a groundbreaking discovery.
- **Vera Rubin (1928–2016)**:
  - An American astronomer who provided critical evidence for the existence of dark matter through her pioneering observations of the rotation of galaxies.
- **Marie Curie (1867-1934)**:
  - Her groundbreaking work on radioactivity and atomic structure laid the foundation for many of the key concepts and measurements that would become integral to the development of quantum mechanics. She was the first woman to receive a Nobel Prize and the only woman to receive two!

Women in STEM 2

- **Jocelyn Bell Burnell (1943–Present)**:
  - A British astrophysicist who discovered pulsars (rotating neutron stars) during her doctoral research in radio astronomy in 1967. Her discovery revolutionized our understanding of celestial objects.
- **Andrea Ghez (1965–Present)**:
  - An American astrophysicist who was awarded the Nobel Prize in Physics in 2020 for her work on the discovery of a supermassive black hole at the center of our galaxy. She is the 4th woman to receive the Nobel Prize in Physics.
- **Harvard Computers (late 1800s-early 1900s)**:
  - A group of women who worked at the Harvard College Observatory under Edward Charles Pickering. These women, often referred to as "computers," made significant contributions to astronomy by conducting painstaking and detailed work in data analysis and cataloging of celestial objects. They worked for mere cents. Some names have been mentioned already. Others include Scottish astronomer Williamina Fleming and American astronomers Annie Jump Cannon, Antonia Maury, and Henrietta Swope.
- **Nancy Grace Roman (1925-2018)**:
  - NASA's first-ever chief of astronomer and "Mother of Hubble." She played a pioneering role in the launch of the telescope. I mentioned that there's a future telescope named after her.
- **Katherine Johnson (1918-2020)**:
  - She was a human computer in a group of female mathematicians who performed complex calculations by hand for various engineering and scientific projects. Her work allowed the Apollo 11 astronauts to land on the Moon successfully in 1969. John Glenn, the first American to orbit Earth, said regarding the numbers, "If she says they're good, then I am ready to go."
- **Kalpana Chawla (1961-2003,** no I wasn't named after her):
  - She was the first Indian woman in space who made her first spaceflight as a mission specialist on Space Shuttle Columbia's STS-87 mission in 1997. She conducted experiments in the Spacehab module during this mission and participated in a spacewalk. She died during the reentry of the Space Shuttle Columbia.

I can't stress enough how these are only SOME of the remarkable women whom many have never heard of.

Solar System

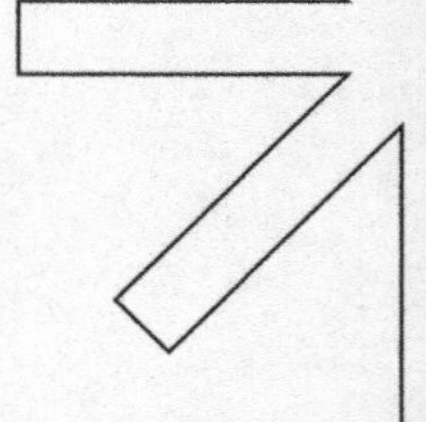

Solar System Basics

Obviously, the above picture is not to scale distance. The Sun has a much more considerable gravitational influence than most realize. It extends well beyond Neptune, the Kuiper Belt, and Heliopause. It extends to the Oort Cloud.

To get as basic as possible, our Sun is a star. It has acquired enough mass to squish elements together into heavier ones and release a crap ton of energy. Planets can't do this. We only see them because they reflect the light coming from the Sun. They aren't self-luminous.

Earth is the third planet from the Sun, and on average, it's about 93 million miles (149,668,992 km) away from it. I say average because all eight planets (we'll get to Pluto later) have varying elliptical orbits. How elliptical an orbit is is known as "eccentricity." Since they're not perfect circles, there are times when the planets are closer to the Sun and times when they are further away. That's why we average their distances.

The Earth-Sun distance of 93 million miles is called one astronomical unit or AU. When it comes to the ENTIRE solar system, using AUs makes the most damn sense. Miles are too damn small, and light years are too damn big. A light year is the distance light travels in a year, which is 5.8 trillion miles (9.3 trillion km).

Defining the edge of the solar system is complicated. It depends on whom you ask. It could be just to the planets, or to the heliopause, or through the Oort Cloud. Regardless, the Sun is indeed the powerhouse in our system, as it makes up 99.8% of the solar system's mass!

SS Structure

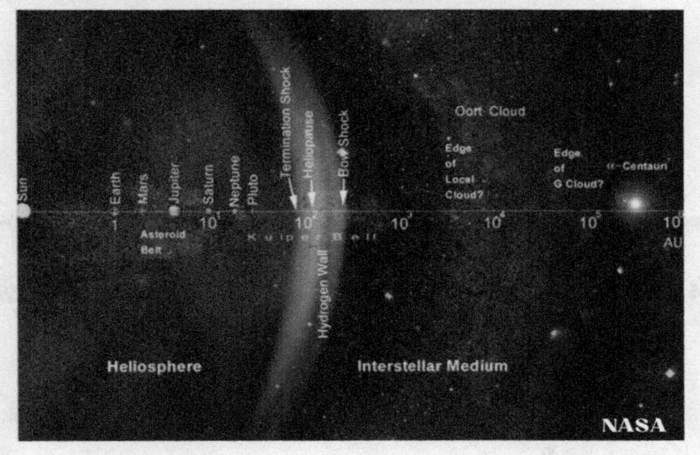

*Note that the above scale is logarithmic, so the distances increase significantly between each one!

The structure of our solar system goes like this.

- Sun at the center (duh).
- The inner planets: Mercury, Venus, Earth, and Mars.
- The Asteroid Belt.
- The gas giants: Jupiter, Saturn, Uranus (lol), and Neptune.
- Pluto, which intersects the Kuiper Belt Objects.
- The Kuiper Belt: a circumstellar disc of icy objects beyond Neptune.
- Heliopause: where the Sun's magnetic field loses influence.
- Interstellar space.
- The Oort Cloud, where the Sun's gravity loses influence at its edge.

Planets don't fall into the Sun because they are moving too fast in their direction of travel (tangential direction). This velocity (speed in a given direction) is enough to overcome falling towards the Sun.

Sun to Mercury distance: 42.7 million miles (68.7 m km), 0.4 AU
Sun to Neptune distance: 2.78 billion miles (4.47 b km), 30 AU
Sun to Heliopause distance: 123 AU
Sun to Oort Cloud distance: 2000 - 100,000 AU! (The range is where the Oort Cloud begins and ends.)

These are averages and approximations. When you're talking about scales as large as these, absolute precision isn't necessary or possible. This is basic astronomy, not physics!

Discoveries

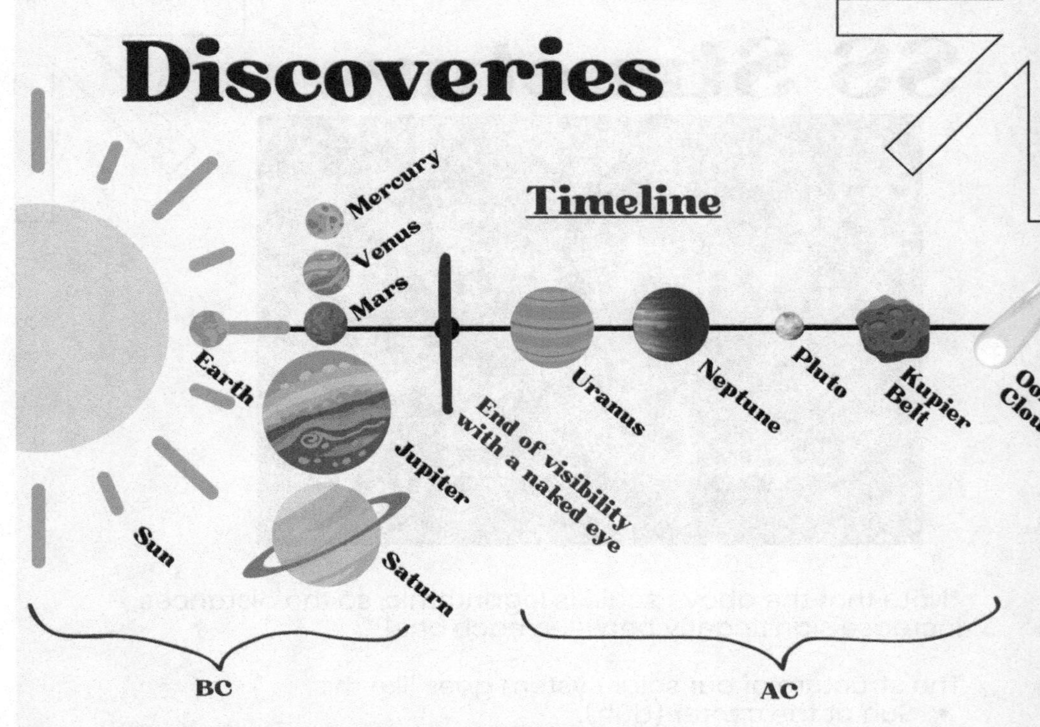

- Earth: discovered by those that live here.
- Sun: discovered by those that live on Earth.

The planets up to Saturn looked like stars that moved differently in relation to the background stars.

- Mercury: the earliest known mention is from the Sumerians around 3000 BCE and observed by all sentient beings with eyes. It was first seen through telescopes in the 1600s by Galileo.
- Venus: ditto.
- Mars: ditto.
- Jupiter: ditto.
- Saturn: ditto.

End of visibility with the naked eye. (Need telescopes beyond this.)

- Uranus (lol): discovered by William Herschel in 1781 through a telescope.
- Neptune: discovered by three British astronomers, Urbain Le Verrier, Johann Gottfried Galle, and John Couch Adams, in 1846 through math and telescopes.
- Pluto: speculated to exist by Percival Lowell but discovered by Clyde Tombaugh in 1930.
- Kuiper Belt: speculated to exist by Gerard Kuiper in 1951 but discovered by American astronomers Dave Jewitt and Jane Luu in 1992 (that rhymes).
- Oort Cloud: existence predicted by Dutch astronomer Jan Oort in the 1950s based on studies of long-period comets.

Formation

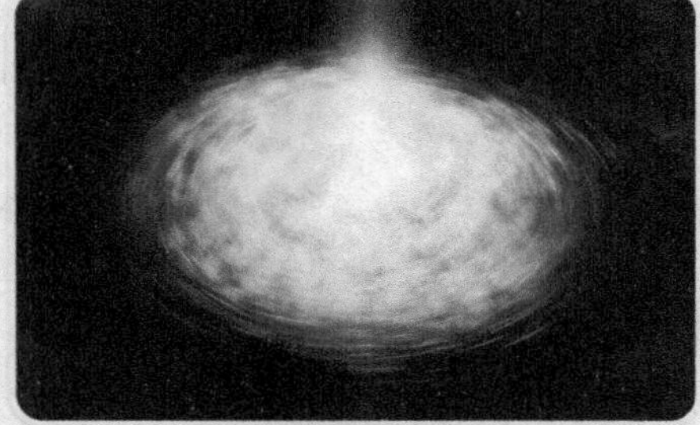

Our solar system began to form around 4.5 billion years ago. We know this is the general time frame because scientists use radiometric dating techniques, such as uranium-lead dating and lead-lead dating to determine the age of certain minerals within the oldest meteorites we find, which have the most extreme lead ratio. We can even date Moon rocks. The Moon hasn't undergone the same geological processes that Earth's rocks have endured. Therefore, they've been well preserved since the Moon's formation.

From death comes life. Our solar system more than likely formed when a nearby star died, went supernova, and sent a shockwave that set into motion the collapse of our gigantic cloud of gas and dust, thus forming a solar nebula, a swirling disk of material. Once the Sun formed and all the leftover material flattened out, the planets would eventually form from this protoplanetary disk.

Our Sun is considered a third-generation star, meaning two stars preceded it. This is theorized due to the amount of heavier elements in the Sun. These wouldn't exist without the previous stars creating them in death.

The inner planets (Mercury-Mars) are rocky because only the materials that could withstand the Sun's extreme heat during formation could last. This would evaporate the Hydrogen and Helium from close planets, while further away ones, the gas giants (Jupiter-Neptune), could keep their gases. The last two planets, Uranus (lol) and Neptune, are ice giants as they have more atmospheric water and ice-forming molecules that freeze since they're even further from the Sun.

Just imagine how fucking chaotic this period was!

FUN FACT

The way the planets formed led to their characteristics. All planets except for Venus and Mars have a planet-wide magnetic field. Earth's is quite strong and caused by molten iron and nickel movement in its core. The magnetic field is like a giant shield that protects the planet from harmful solar radiation. We wouldn't exist without it!

The Sun

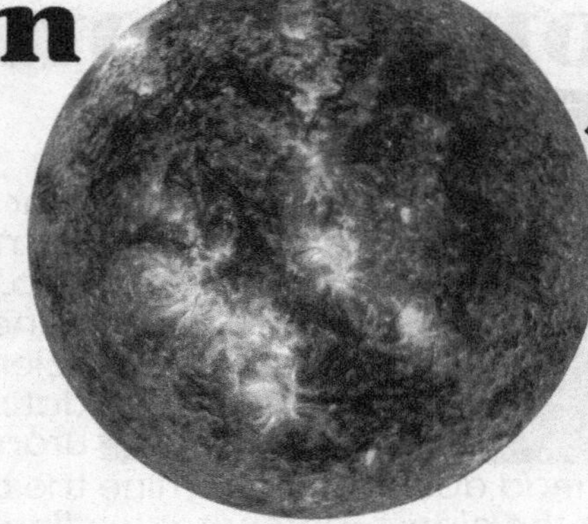

Jupiter to Sun size.

Some fun Sun facts.

- **Life**:
  - Halfway through its ten billion-year life.
  - Currently fusing hydrogen into helium in the core.
  - In 5 billion years, it will run out of hydrogen and begin to fuse helium, marking the beginning of the end of its life.
- **Composition**:
  - Primarily composed of 74% hydrogen, 24% helium, and < 2% of other elements.
  - Magnetic field is 2x stronger than Earth's and extends well beyond the furthest planet.
- **Size**:
  - Diameter is about 864,337 miles (1,391,015 km).
  - About 109 times larger than Earth.
  - 1.3 million Earths can fit inside of it (volume).
  - Considered to be a yellow dwarf as it fits in one of the smaller steller classes (more on this later).
- **Energy**:
  - Currently produces an estimated 383 billion trillion watts (3.8×10^{26} watts) of energy every second.
  - Life on Earth depends on energy from the Sun. Plants need it for fuel, and we need the plants for fuel.
  - It's the best source of vitamin D.
- **Sun-studying spacecraft**:
  - Parker Solar Probe, Solar Dynamics Observatory, Solar and Heliospheric Observatory, Solar Orbiter, Advanced Composition Explorer, and Interface Region Imaging Spectrograph.
- I hate the heat and being in the Sun.

The Parker Solar Probe is the fastest man-made object, traveling at 430,000 mph (692,017.92 km/hr)!

The Hottest Star

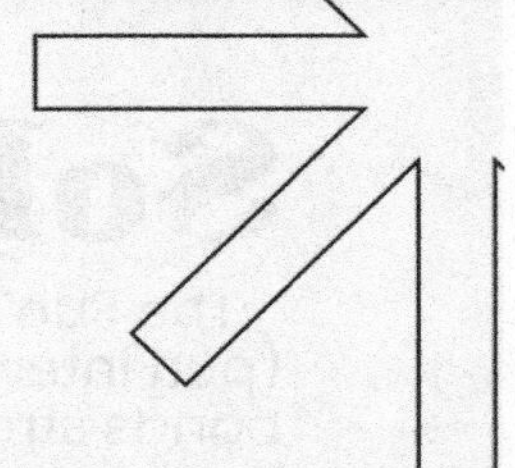

The Sun is our solar system's hottest (and only) star. Sorry, Hollywood. But unlike some Hollywood stars who crave attention to stay relevant, the Sun likes to be secretive.

Temperatures:
- **The core**, where nuclear fusion takes place is 27 million° Fahrenheit (15 m° Celsius).
- **The radiative zone**, where energy created in the core is transported outward as photons, ranges from 27 million - 3.6 million°F (15 m - 2 m°C) moving outward.
- **The convection zone**, where energy is transported through plasma, ranges from 3.6 million to 10,832°F (2 m - 6k°C) as you move to the surface.
- **The photosphere,** the visible part where light is released into space, is 9,932°F (5,500°C).

Now entering the Sun's atmosphere.
- **The chromosphere**, a thin plasma layer under the upper atmosphere, can reach 36,000°F (20k°C).
- **The transition region**, a thin layer of atmosphere ranges in temperatures from about 36,032 to 1.8 million°F (20k - 1 m°C).
- **The corona**, the outermost layer of the Sun with low density, ranges in temperature from about 1.8 to 5.4 million°F (1 m - 3 m°C).

As you leave the core to the surface the temps decrease. As you leave the photosphere (surface) to the atmosphere the temps increase. Scientists aren't entirely sure why that is!

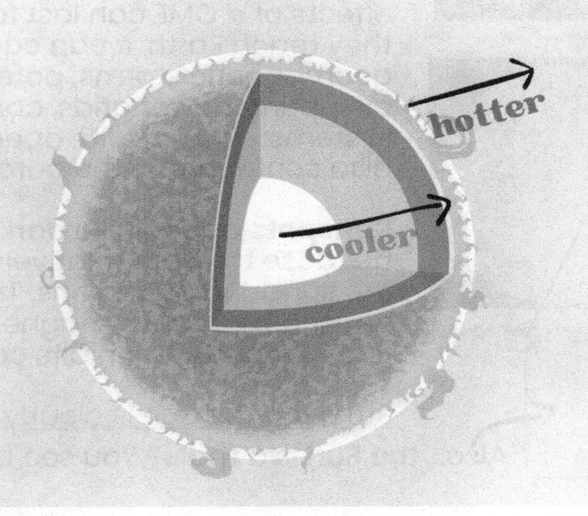

Solar Activity

The Sun's magnetic field is a hot mess (pun intended). Think of all these rubber bands stretching, twisting, and folding back in on themselves. Now imagine that they cover the Sun and are being stretched and snapped back. They cause all of the Sun's activity.

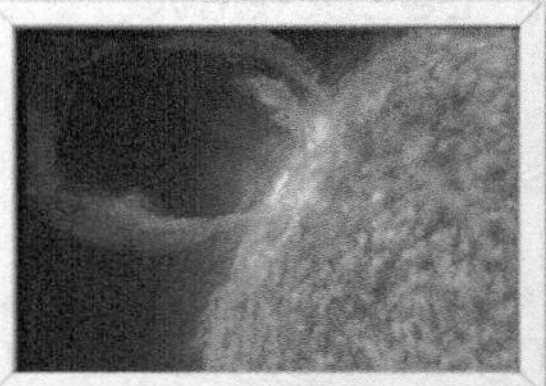

Prominences are large, bright features extending outward from the Sun's surface, often appearing as loops or plumes held in place by magnetic fields.

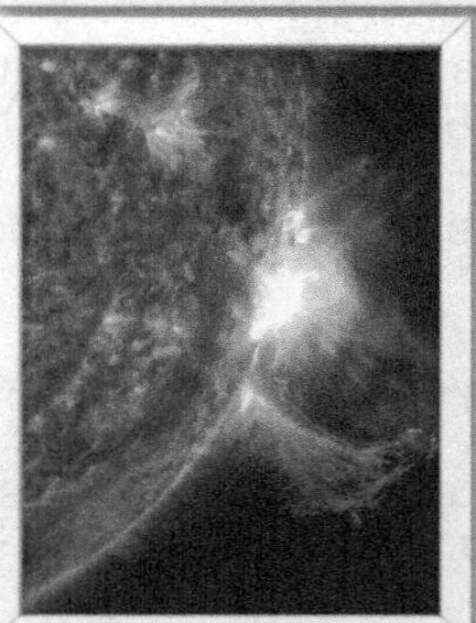

Solar flares are brief eruptions of intense high-energy radiation from the Sun's surface. They appear as bright patches, can last from minutes to hours, and travel at the speed of light. They can disrupt radio communications, GPS systems, and power grids, and pose radiation risks to astronauts.

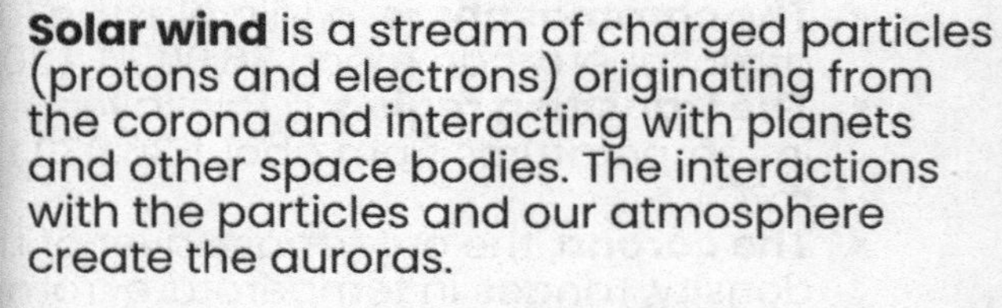

Solar wind is a stream of charged particles (protons and electrons) originating from the corona and interacting with planets and other space bodies. The interactions with the particles and our atmosphere create the auroras.

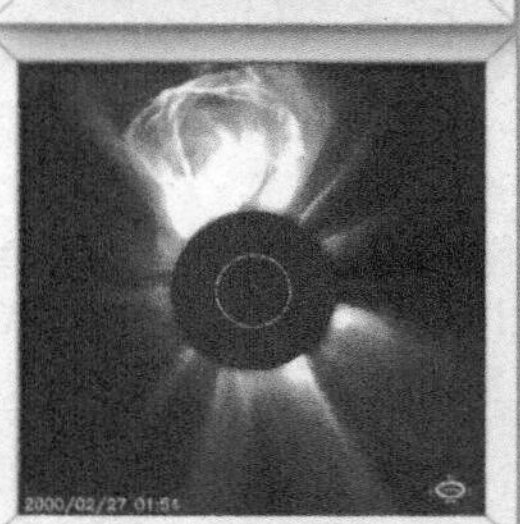

Coronal Mass Ejections (CMEs) are massive bursts of solar wind and magnetic fields released into space. They can release up to 10 billion tons of solar material and travel at speeds > 1 million mph (1.6 million km/hr). The effects of a CME can last for days. When they reach Earth, it can cause geomagnetic storms, potentially disrupting power grids, communication systems, and satellite operations. CMEs also contribute to the auroral displays.

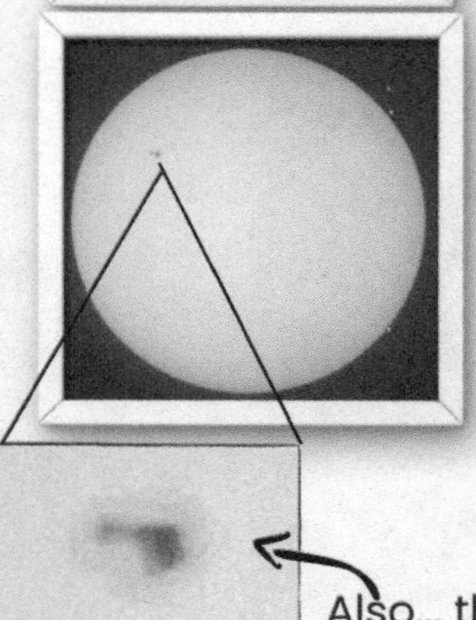

Sunspots appear like dark patches because they are relatively cooler than the surrounding regions. They are caused by intense magnetic activity that temporarily inhibits convection.

<u>Studying the Sun is clearly important!</u>

Also... the Sun is a male? You see it, right?

Rocky Planets

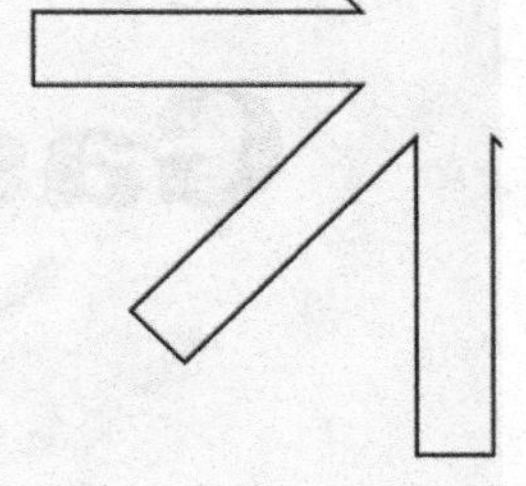

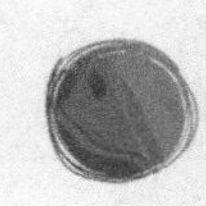   

The first four planets are rocky or terrestrial, meaning they have a solid surface.

- **Mercury**: the closest planet to the Sun with an average distance of ~ 30 million miles (48 m km) away from it.
  - Diameter is about 3,032 miles.
  - 18 Mercurys would equal the mass of 1 Earth.
  - Iron core makes up 70% of its entire body.
  - Very thin atmosphere of hydrogen and helium.
  - Temps range from day 800°F - night 290°F (426 - 143°C).
  - Very weak magnetic field.
  - Rotation is longer than orbit, 176:88 Earth days.
  - Roman name for the Greek messenger god, Hermes.
- **Venus**: average distance to the Sun ~ 67.42 m miles (107 m km).
  - Diameter is about 7,521 miles (12,103 km).
  - 1.2 Venuses would equal the mass of 1 Earth.
  - Mantle made of silicate rocks and an iron-nickel core.
  - A very thick atmosphere primarily composed of CO_2 creates a strong greenhouse effect.
  - It's the hottest planet with temps around 864°F (452°C).
  - Volcanic surface with plains, craters, and mountains.
  - Only planet to spin clockwise. It takes longer to rotate than orbit, 243:225 Earth days.
  - Very bright in the sky and was given the Roman name for the Greek goddess of love, Aphrodite.
- **Earth**: average distance to the Sun ~ 93 million miles.
  - Diameter is about 7,917 miles (12,741 km).
  - A solid iron-nickel core, a mantle made of silicate rocks, and a thin outer crust.
  - Atmosphere is 78% nitrogen and 21% oxygen.
  - Geological activity due to the motion of tectonic plates.
  - Liquid water covers 71% of the planet.
- **Mars**: average distance to the Sun ~ 142 m miles (228 m km).
  - Diameter is 4,212 miles (6,778 km).
  - 9 Mars would equal the mass of 1 Earth.
  - Iron core but very weak magnetic field.
  - Very thin CO_2 atmosphere.
  - Temps can range from 70°F to -225°F (21 to - 142°C).
  - Has the largest canyon, Valles Marenaris, and the largest volcano, Olympus Mons in the solar system.
  - One day is 24 hours and 37 minutes.
  - There's a potential for ice water underground.
  - Entirely inhabited by robots!
  - Roman name for the Greek god of war, Ares.

Gas Giants

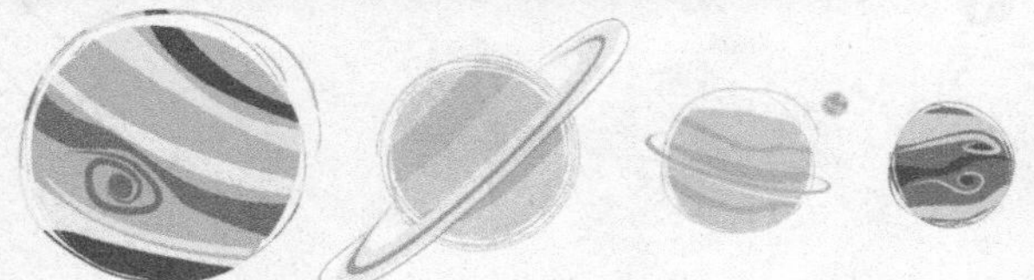

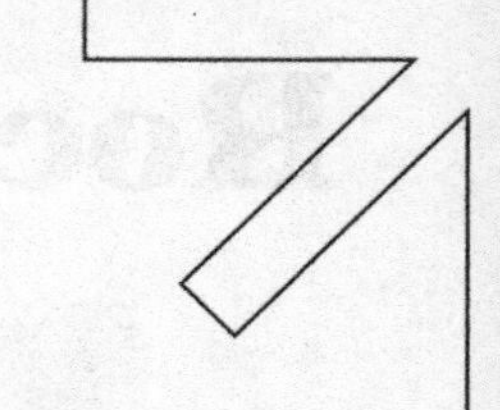

The last four planets are giants made out of gas that swirl around a solid core. All have rings.

- **Jupiter**: the largest planet with an average distance ~ 484 m miles (778 m km) from the Sun.
  - Diameter is about 86,881 miles (139,821 km).
  - 1 Jupiter is about the mass of 318 Earths.
  - Primarily composed of 75% hydrogen and 24% helium.
  - The strongest magnetic field and largest atmosphere.
  - Great Red Spot is a storm bigger than Earth.
  - Day is 10 hours, year is 12 years.
  - Roman name of the Greek king of gods, Zeus.
- **Saturn**: has the largest ring system with an average distance ~ 886 m miles (1.4 b km) from the Sun.
  - Diameter is about 72,366 miles (116,461 km).
  - 1 Saturn is about the mass of 95 Earths.
  - The lowest density of any planet as it would float in a (big enough) bathtub.
  - Primarily composed of 96% hydrogen and 3% helium.
  - Layers include gaseous, liquid, and metallic hydrogen around a rock and metal core.
  - Strong magnetic field and atmosphere.
  - Rings are ice particles that might be a broken-up moon.
  - Roman name for Zeus's father, the Greek titan, Cronos.
- **Uranus** (lol): avg. distance to the Sun ~ 1.8 b miles (2.9 b km).
  - Diameter of this ice giant is about 31,518 miles (50,723 km).
  - 1 Uranus is about the mass of 14.5 Earths.
  - Thick atmosphere comprises 83% hydrogen, 15% helium, and 2% methane, which surrounds the larger mantle of water, ammonia, and methane.
  - Gets its blue-green color from methane in atmosphere.
  - Each side faces the Sun for 21 years during its orbit.
  - Originally named George after King George III.
  - Later given the Roman name for the Greek entity of the Heaven, Ouranos. (I'm not even joking.)
- **Neptune**: coldest and furthest planet from the Sun with an average distance of 2.8 b miles (4.5 b km).
  - Diameter of this ice giant is ~ 30,598 miles (49,242 km).
  - 1 Neptune is about the mass of 17 Earths.
  - Atmosphere is composed of 80% hydrogen, 19% helium, and 1% methane, giving it its blue color.
  - Has a dark storm spot like Jupiter.
  - Fastest winds and likely rains "diamonds" deep within.
  - Roman name for the Greek god of the sea, Poseidon.

Kepler's Laws

Unlike the Earth, the planets' orbital plane around the Sun is relatively flat. This is known as the ecliptic and is also the Sun's apparent path over a year. All eight planets orbit within 3 degrees of this plane. Pluto's orbit deviates to ~ 17 degrees, and other Kuiper Belt objects deviate up to 30 degrees. All planets have elliptical orbits. The more eccentric (0-1), the greater the ellipse. Mercury is the most elliptical, while Venus is the most circular.

Planet Eccentricities

Planet	Eccentricity
Mercury	0.205
Venus	0.007
Earth	0.017
Mars	0.094
Jupiter	0.049
Saturn	0.057
Uranus	0.046
Neptune	0.011
Pluto	0.244

Basic Anatomy of an Ellipse

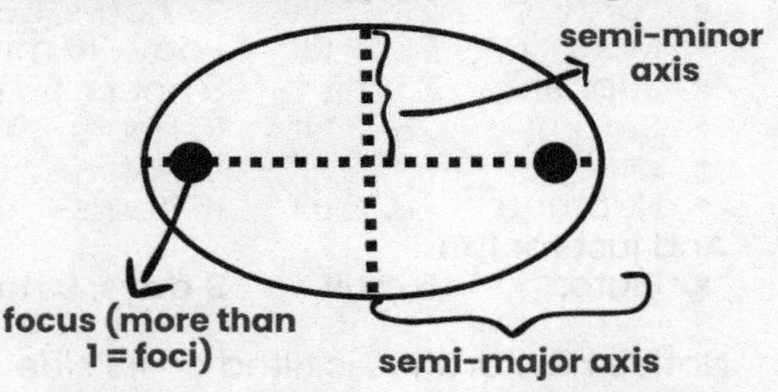

In the 1600s, German mathematician Johannes Kepler created his three laws to describe the orbits of the planets. To sum up:

1) The Sun is located at a focus, and the planets orbit in ellipses, so their distances to the Sun constantly change as they orbit.

2) Planets don't orbit the Sun at a constant speed. Their speeds increase when they are closer to the Sun and decrease when they are further away, but they still "sweep" the same amount of area in the same amount of time. Both blue areas cover the same amount of space. When the planet is closest to the Sun, it's known as perihelion; when it's farthest away, it's known as aphelion.

3) The time it takes to orbit rapidly increases with distance. Thus, there is a relationship between the orbital period and the semi-major axis, the planet's average distance from the Sun. The square of the orbital period is directly proportional to the cube of the semi-major axis. As in, if you know how long it takes for a planet to orbit, you can find its distance. Kepler didn't know about gravity and didn't know that this law set the basis for Newton to explain the force behind it eventually.

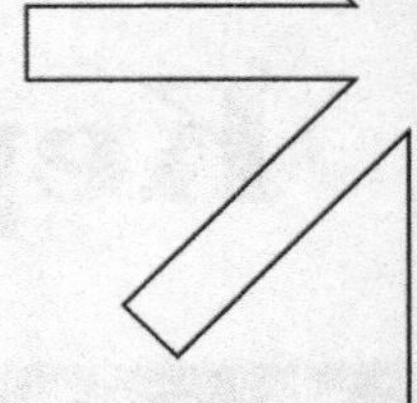

Uranus's Tilt (lol)

It's not just Uranus (lol), but all planets have a tilt, and they all orbit counterclockwise around the Sun. The axis of rotation, an imaginary line that each planet rotates around, is relative to the ecliptic. That's how a planet's tilt is determined and how we measure Earth's tilt to be 23.4°. Here are all the planets' tilts, length of days, and orbital periods around the Sun measured in Earth days.

Planet	Tilt	Day Length	Orbit
• Mercury:	0° tilt	176 days	88 days
Rotates on its axis in 58 days, but sunrise comes in 176 days.			
• Venus:	177.3° tilt	243 days, 26 minutes	225 days
• Earth:	23.4° tilt	23 hours, 56 minutes	365 days
• Mars:	25.2° tilt	1 day, 36 minutes	687 days
• Jupiter:	3.1° tilt	9 hours, 55 minutes	4,333 days
• Saturn:	26.7° tilt	10 hours, 40 minutes	10,759 days
• Uranus:	97.8° tilt	17 hours, 14 minutes	30,687 days
• Neptune:	28.3° tilt	16 hours	60,190 days
And just for fun...			
• Pluto:	57° tilt	9 days, 9 hours	90,520 days

Notice how Uranus is tilted on its side (better get that checked), and Venus is pretty much upside down. Earth spins west to east, but Venus spins east to west. Therefore, it's considered upside down *relative to us*.

The tilts were more than likely created during the chaos of the solar system formation as the newly forming planets were getting hit all the time. On Earth (and on all the planets), the tilt is one of the most important contributors to having seasons. When we tilt toward the Sun, we get warmer seasons, and when we tilt away, well, I'm sure you can figure that part out.
Also... winter > summer. Comet me, bro!

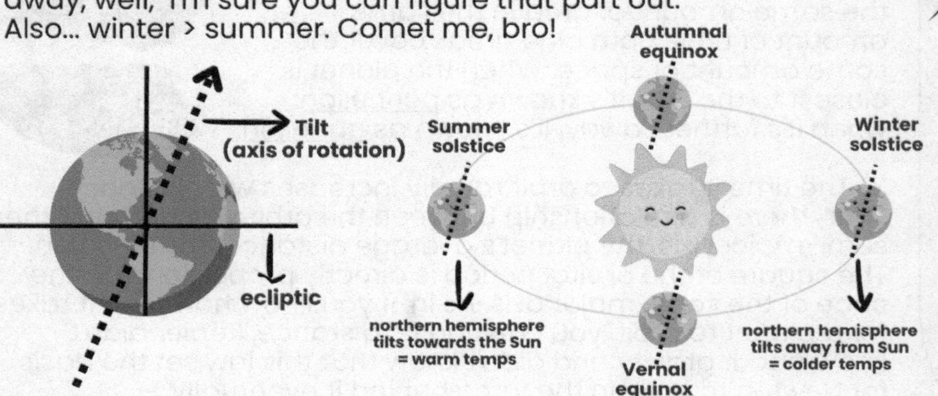

Spacecraft

That have visited the planets.

Mercury:

- **Mariner 10 (NASA)**: Conducted multiple flybys of Mercury in the mid-1970s, providing the first close-up images.
- **MESSENGER (NASA)**: Orbited Mercury from 2011 to 2015, studying its surface, composition, and magnetic field.

Venus:

- **Mariner 2 (NASA)**: The first successful mission that conducted a flyby in 1962 and provided valuable data about the planet's atmosphere and surface temperature.
- **Venera Series (Soviet Union)**: A series of successful missions, including landers and orbiters, that provided extensive data about Venus' atmosphere and surface. Venera 13 set the record for lasting just over 2 hours on the volatile surface.
- **Magellan (NASA)**: Mapped the surface using radar, providing detailed topographical information.

Mars:

- **Viking 1 and 2 (NASA)**: The first successful landers, arriving in 1976, conducted biological experiments and analyzed the Martian surface.
- **Mars Rovers (Sojourner, Spirit, Opportunity, Curiosity, Perseverance) (NASA)**: Various rovers have explored the Martian surface, studying geology, climate, and potential for past or present life.
- **InSight (NASA)**: Studied its interior and seismic activity.

Jupiter and Saturn:

- **Pioneer 10 and Pioneer 11 (NASA)**: Conducted flybys of Jupiter, providing valuable data about its atmosphere and magnetic field. Only Pioneer 11 for Saturn.
- **Voyager 1 and 2 (NASA)**: Conducted flybys providing extensive data on the planets, their moons, Jupiter's magnetosphere, and Saturn's rings.
- **Galileo (NASA)**: Studied Jupiter's atmosphere, magnetosphere, and moons, including discovering a subsurface ocean on Europa.
- **Cassini-Huygens (NASA/ESA/ASI)**: Studied Saturn, its rings, and diverse moon system, including the Huygens probe landing on Titan.

Uranus (lol) and Neptune:

- **Voyager 2**: Conducted a flyby and collected data on their atmospheres, rings, and moons, and gave us images.

Dwarf Planets:

- **New Horizons (NASA)**: Conducted a historic flyby of Pluto and its moons, providing detailed images and data.
- **Dawn (NASA)**: Explored the dwarf planet Ceres, studying its composition and surface features.

ISS

The International Space Station (ISS) is a remarkable and collaborative effort involving space agencies from multiple countries to create a habitable space station in low Earth orbit. It serves as a microgravity and space environment research laboratory where scientific research is conducted in astrobiology, astronomy, meteorology, physics, and others.

- Agencies include NASA, Roscosmos, ESA, JAXA, and CSA.
- The project began in the late 1990s, with the first module, Zarya, launched in 1998.
- It orbits Earth at an average altitude of ~ 260 miles (418 km) and travels about 17,500 mph (28,163 km/hr).
- It completes an orbit around Earth every 90 minutes, and is visible to the naked eye when it passes over at night!
- It hosts a crew of 3-6 international astronauts at all times who conduct scientific experiments and research.
- So far, it has had over 200 astronauts from 21 countries.
- The unique microgravity environment allows for experiments that cannot be conducted on Earth.
- Significant discoveries and research include the effects of microgravity on the human body, potential treatments for osteoporosis, cardiovascular health, plant growth, crystallization experiments that lead to drug development, space radiation, and Earth observations.
- Sadly, the ISS is set to end in 2030 due to continuous stress on the spacecraft. It will then crash into the Pacific Ocean.
- The ISS is a testament to international cooperation in space exploration. It fosters collaboration and peaceful relations among nations despite geopolitical differences. Science brings people together!

Since 2000, not all humans have been on Earth because the ISS has continuously been inhabited since then!

Moons

It blows my mind that many people don't know wtf a "moon" is. Remember that viral QVC video? *face-palm*

A <u>moon</u> is a natural object that revolves around a planet or something that isn't a star. They're natural satellites.

All the planets except for Mercury and Venus have moons.

Mars has two oddly shaped moons, Phobos and Deimos. Because they're weird, scientists aren't sure how they ended up around Mars. Were they captured asteroids? Were they blown off from an early Mars? Did they form from a ring around Mars? We're not sure. Scientists predict that Phobos will crash into Mars or break up and form a ring in about 50 million years.

All gas giants have dozens of moons. Jupiter and Saturn are always competing for the title of most moons. At the time of writing this, Saturn retook the title with 145 moons.

- Notable Saturn moons:
  - Titan: icy and the 2nd largest moon in the solar system.
  - Enceladus: icy moon with a subsurface ocean.
  - Rhea: icy and only one with a thin oxygen atmosphere.
  - Iapetus: icy and has bizarre dark and bright sides.
- Jupiter, 95 moons including the 4 Galilean Moons:
  - Io: most volcanically active place in the solar system.
  - Europa: icy with an intruiging subsufrace ocean.
  - Ganymede: largest moon and bigger than Mercury.
  - Callisto: most heavily cratered object in the solar system.
- Uranus (lol), 27 moons, notable moons all thought to consist of water ice and silicate rock:
  - Miranda: smallest and innermost of the five.
  - Ariel: brightest surface of the five.
  - Umbriel: darkest surface of the five.
  - Titania: largest of all 27 and geologically active.
  - Oberon: heavily cratered and outermost of the five.
- Neptune, 14 moons:
  - Triton: largest of 14 and the only moon in the solar system to orbit in the opposite direction of its planet's rotation.

Pluto has 5 moons: Charon, Nix, Styx, Kerberos, and Hydra. Pluto and its main moon, Charon orbit each other. Some say this should make them a binary planet system.

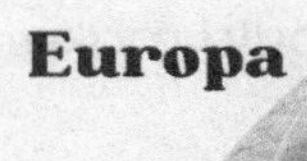

Europa

Our Moon

the only moon named Moon

- Orbits Earth in 27.3 days.
- Average distance to Earth is 238,885 miles (383,000 km).
- 1/4th of Earth's size and 1/6th of its gravity.
- Goes through phases: new moon, first quarter, full moon, and last quarter.
- Heavily cratered and has dark, flat plains known as Maria (Latin for sea), as it was thought they were bodies of water.
- Has a very thin atmosphere known as an exosphere, composed of scattered atoms and molecules.
- Has water ice in certain parts.

It's theorized that our Moon formed after a Mars-sized object named Theia crashed into the early Earth. This violent collision threw out a ton of material, which reformed into our Moon. The Moon was necessary for life to form on Earth as it stabilized our erratic rotation.

Is it true that the same side always faces the Earth? Yes, but contrary to popular belief, it's not because the Moon *doesn't* spin; it's because it *does*! The Moon and Earth are tidally locked, which means it takes as long for the Moon to rotate once on its axis as it does to orbit the Earth once. This is why we always see the same side. It seems counterintuitive, but we would see the other side if the Moon didn't spin on its axis. Put something on the floor and walk around it in a circle. Walk once where your toes always face the object, and once where you don't change angles, so eventually your heels face it. Technically, you are spinning once on your axis in the first circle because your angle changes. That's what our Moon does.

The Moon doesn't exactly revolve around us. It and the Earth revolve around a center of mass known as the barycenter. This is where the Earth's and the Moon's combined mass (or any two or more objects) is evenly distributed. This point is not at the center of Earth but 2,900 miles (4,667 km) above it, or at 75% of Earth's radius. Both bodies revolve around this point!

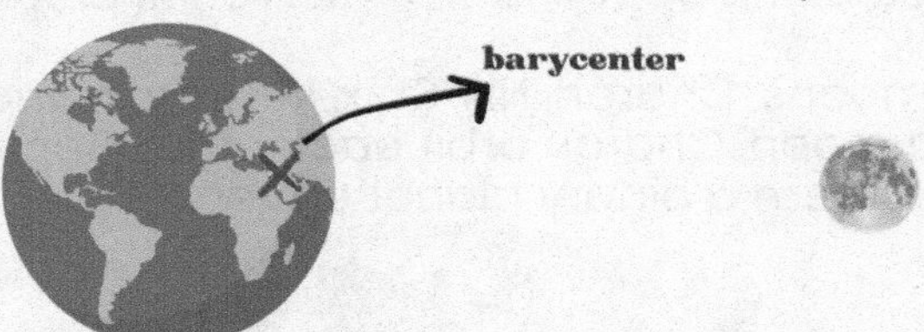

Like with Pluto and Charon, some argue that this should make the Earth and Moon a binary planet system!

The Moon is the furthest humans have reached. Speaking of...

To the Moon

Notable moon missions. Yes, we landed on the Moon.

- **Apollo Program (NASA)**:
    - **Apollo 11 (1969)**: 1st time humans landed on the Moon!
    - **Apollo 12, 14, 15, 16, 17**: Continued to land astronauts on the Moon, conducted scientific experiments, explored the lunar surface, and advanced our understanding of the Moon's geology and composition.
- **Luna Program (Soviet Union)**:
    - **Luna 2 (1959)**: 1st human-made object to reach the Moon. It crash-landed on the Moon's surface, providing the first direct solar wind measurements.
    - **Luna 9 (1966)**: 1st successful soft landing on the Moon and the first mission to transmit photographs of the lunar surface back to Earth.
    - **Luna 16, 20, 24**: Sample return missions that collected lunar soil and brought it back to Earth.
- **Surveyor Program (NASA)**:
    - **Surveyor 1-7 (1966-1968)**: A series of robotic missions that successfully soft-landed on the Moon, providing valuable data for the Apollo program of safe landing sites.
- **Ranger Program (NASA)**:
    - **Ranger 7-9 (1964-1965)**: These were intended to impact the Moon and transmit high-resolution images of the lunar surface before impact.
- **Chang'e Program (China)**:
    - **Chang'e 3 (2013)**: China's first successful soft landing on the Moon with the deployment of a rover named Yutu (Jade Rabbit). It conducted experiments and explored the lunar surface.
    - **Chang'e 4 (2019)**: China's mission to explore the far side of the Moon, making a historic landing and deploying the Yutu 2 rover.
- **Lunar Reconnaissance Orbiter (NASA)**:
    - **LRO (2009)**: An ongoing mission orbiting the Moon, mapping its surface, studying its radiation environment, and providing data for future human exploration. You can see evidence of our past visits in LRO images. (Not that it matters to conspiracy nuts.)
- **Chandrayaan program (India)**:
    - **Chandrayaan-3 (2023):** India's 1st successful moon landing and the first spacecraft to land on the Moon's south pole.
- **Artemis Program (NASA)**:
    - An upcoming series of crewed missions to return humans to the Moon. The Artemis program aims to land the first woman and the next man on the Moon and establish a sustainable lunar presence.

The Fallen

Never forget the brave men and women who lost their lives in pursuit of space exploration.

- **Soyuz 1 (Soviet Union, 1967)**: A parachute failure brought the spacecraft crashing down from a height of 4.5 miles (7.2 km). Vladimir Komarov (Commander) was killed instantly.
- **x-15 Flight 3-65-97 (USA, 1967)**: As a part of the x-plane series experimental aircraft, the x-15 was a hypersonic rocket-powered aircraft. Michael J. Adams (Pilot) died when his aircraft broke apart after just over 10 minutes of flying. He was awarded astronaut wings posthumously.
- **Apollo 1 (USA, 1967)**: A fire swept through the command module during a launch rehearsal test and killed everyone on board:
    - Virgil "Gus" Grissom (Command Pilot)
    - Edward H. White II (Senior Pilot)
    - Roger B. Chaffee (Pilot)
- **Soyuz 11 (Soviet Union, 1971)**: The crew capsule depressurized during preparations for reentry, killing everyone on board:
    - Georgi Dobrovolski (Commander)
    - Viktor Patsayev (Research Engineer)
    - Vladislav Volkov (Research Engineer)
- **Space Shuttle Challenger (USA, 1986)**: Broke apart 73 seconds into the flight, killing everyone on board:
    - Francis R. Scobee (Commander)
    - Michael J. Smith (Pilot)
    - Ronald McNair (Mission Specialist)
    - Ellison Onizuka (Mission Specialist)
    - Judith Resnik (Mission Specialist)
    - Gregory Jarvis (Payload Specialist)
    - Christa McAuliffe (Payload Specialist, Teacher in Space)
- **Space Shuttle Columbia (USA, 2003)**: reentered Earth's atmosphere, but prior damage to the left wing allowed hot gases to penetrate, leading to the breakup of the shuttle, killing everyone on board:
    - Rick D. Husband (Commander)
    - William C. McCool (Pilot)
    - Michael P. Anderson (Payload Commander)
    - Ilan Ramon (Payload Specialist, first Israeli astronaut)
    - Kalpana Chawla (Mission Specialist)
    - David M. Brown (Mission Specialist)
    - Laurel B. Clark (Mission Specialist)

Gone but not forgotten, we forever honor these brave people!

Eclipses

I urge everyone to see a total solar eclipse!

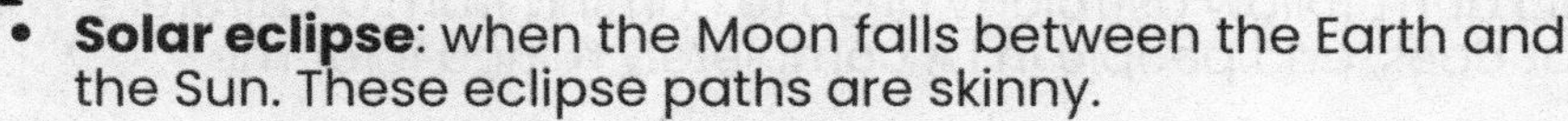

- **Solar eclipse**: when the Moon falls between the Earth and the Sun. These eclipse paths are skinny.

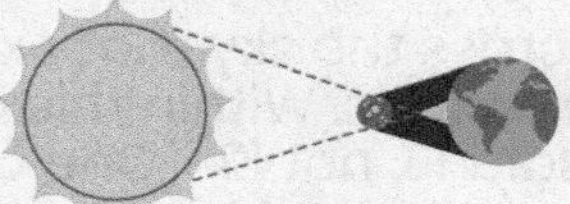

  - **Partial solar**: when the Moon covers only a part of the Sun. They occur on average about twice a year.
  - **Total solar**: when the Moon covers all of the Sun, and no protective eyewear is needed for these minutes of totality. They take place every 18 months or so.
  - **Annular solar**: when the Moon is further away from the Earth in its orbit, it appears smaller. Therefore, there's "less Moon" to completely cover the Sun. These happen every 1-2 years.
  - FUN FACT: Our Moon is the only Moon in the solar system to completely cover the Sun during an eclipse! The ratios just worked out... weird.
  - The next total solar in the US is April 8th, 2024!!

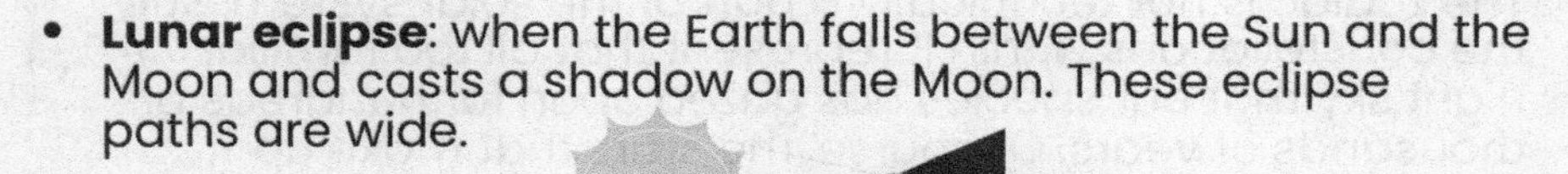

- **Lunar eclipse**: when the Earth falls between the Sun and the Moon and casts a shadow on the Moon. These eclipse paths are wide.

  - The Earth has two parts to its shadow: the outer lighter part, or penumbra, and the inner darker part, or umbra.
  - **Partial lunar**: when the Moon passes through only a part of the umbra due to an imperfect alignment.
  - **Total lunar**: when the Moon is entirely in the Earth's umbra. A "blood moon" can occur because light passing through Earth's atmosphere can reach the Moon's surface, and only longer, redder wavelengths can make it through.
  - There could be up to 3 partial or totals a year.
  - Penumbral lunar: when the Moon passes into the penumbra. We can hardly notice this one as it is in the light part of the shadow.

Zodiac

This is NOT astrology. I'm not shitting on anyone's beliefs, but I don't follow astrology despite coming from a culture that does, and people in LA completely ruined it for me.

The zodiac is a belt across the sky that extends 8 degrees above and below the ecliptic. Within this belt are 13 constellations. Yes, I said 13, not 12! (One of the many reasons I don't follow astrology.) You probably know the 12: Aquarius, Pisces, Aries, Taurus, Gemini, Cancer, Leo, Virgo, Libra, Scorpio, Sagittarius, and Capricorn.

The 13th is between Scorpio and Sagittarius and is called Ophiuchus (the serpent bearer). Why is it not commonplace to acknowledge it in astrology? Beats the hell out of me. It's probably because they don't look into all the facts of what the zodiac is.

The Sun passes in front of each constellation for a certain amount of time. It's not exactly one month each, and even Ophiuchus is behind the Sun for about three weeks.

The zodiac is not technically a part of the solar system. Still, the constellations within it represent that all too familiar night sky that our species has gazed upon for hundreds of thousands of years. Of course, the stars that make up these constellations are moving and at varying distances; therefore, the current patterns won't last forever.

If our species is lucky enough to still be on this planet 1 million years from now, I wonder what those future people will see. But let's be real, our ancestors were probably on crack to have seen what they claimed to see in some of these patterns.

How is this a crab?!

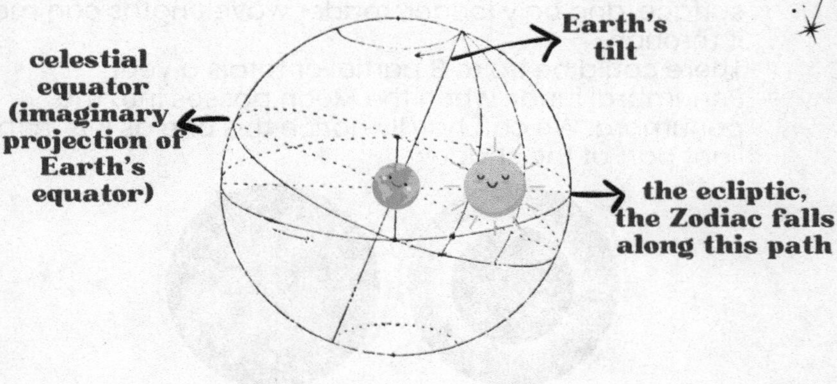

Earth's tilt

celestial equator (imaginary projection of Earth's equator)

the ecliptic, the Zodiac falls along this path

Asteroids, Comets, Meteors, oh my!

- **Asteroids**: the leftover rocky remnants of the formation of the solar system. They reside within the asteroid belt between Mars and Jupiter and in the Kuiper Belt.
    - There are over 1.1 million known asteroids.
    - Over 34k of them are NEOs (near-Earth objects), as they come within 1.3 AU of the Sun.

- **Comets**: cosmic dirty snowballs, meaning they have ice water, dust, and other particles. They comprise four parts: the nucleus, the coma, the dust tail, and the ion tail. Comets originate from the Kuiper Belt and Oort Cloud.
    - The <u>nucleus</u> is the solid body generally a few miles (km) wide.
    - The <u>coma</u> is the freely escaping atmosphere around the nucleus that forms as the comet gets near the Sun.
    - The <u>dust tail</u> is what spews out from this atmosphere. This is what we pass through every time there's a meteor shower. They glow white or yellow in color.
    - The <u>ion tail</u> forms from ultraviolet radiation ejecting electrons from the coma. This tail is different from the dust tail and generally points away from the Sun. They glow bluish in color.

- **Meteors**: are shooting stars. No joke. They cause the streaks we see in the sky. <u>Meteoroids</u> are small fragments of asteroids and comets. <u>Meteors</u> are when these small fragments burn up in our atmosphere. <u>Meteorites</u> are rocks from space that land on the ground.
    - Shooting stars may be bright as hell, but the meteors that create them are only grain-sized!

Comet Neowise and me, July 2020. Comet by @chondrite55.

Armaggedon

Notable asteroid and comet missions.

Believe it or not, we've actually landed on asteroids and comets. Space agencies are the only defense against a possible doomsday from these celestial bodies, one of the MANY ways they help humanity. Contrary to popular (or Hollywood) belief, you don't want to blow up the asteroid. That would fuck things up more as more pieces would rain down on us. Instead, you'd want to nudge it out of the way so its trajectory misses Earth. There are > 2,300 NEOs that are considered potentially hazardous (460 ft, 140 m), and > 850 of those are > 1 km! Those would fuck us up.

It's not just for defense; scientists study these well-preserved bodies to learn about our early solar system, their impact on its evolution and check for the chemical building blocks of life.

- **NASA Missions:**
    - Double Asteroid Redirection Test (DART) successfully impacted asteroid Dimorphos in September 2022, in the agency's first attempt to move an asteroid in space. Dimorphos is a moonlet to asteroid Didymos. It's the world's first planetary defense technology demonstration.
    - Dawn spacecraft launched in 2007 to explore the second largest body in the asteroid belt, asteroid Vesta. Dawn arrived in 2011 and explored for a year before leaving for dwarf planet Ceres, also in the asteroid belt.
    - OSIRIS-REx launched in 2016 and arrived at NEO Bennu in 2018. After some struggle, it collected a sample which it brought back to Earth in September 2023. OSIRIS-REx is now on its way to asteroid Apophis.
        - Bennu is known as "the world's most hazardous asteroid" as it may strike Earth in 2182, which would be catastrophic.
        - Apophis is also a NEO that would be destructive as fuck to Earth. Over time, its impact likelihood has decreased, but it's better to be safe than sorry!
- **ESA Mission:**
    - Rosetta mission's Philae lander landed on comet 67P/Churyumov-Gerasimenko in 2014 and became the first spacecraft to ever land on one. The data collected shows us that comets could have played a role in bringing life's building blocks to Earth. It also flew by asteroid 2867 Steins in 2008 and asteroid Lutetia in 2010.
- **JAXA Missions:**
    - Hayabusa spacecraft landed on asteroid Itokawa in 2005 and brought back a sample in 2010.
    - Hayabusa2 launched in 2014 and arrived at asteroid Ryugu in 2018. It too brought back a sample in December 2020. Scientists are currently studying these samples.

Recipe for Life

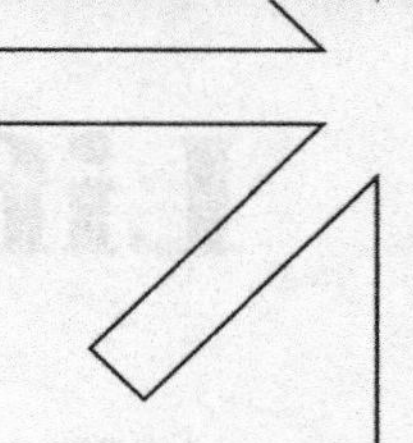

Ingredients

CHNOPS:

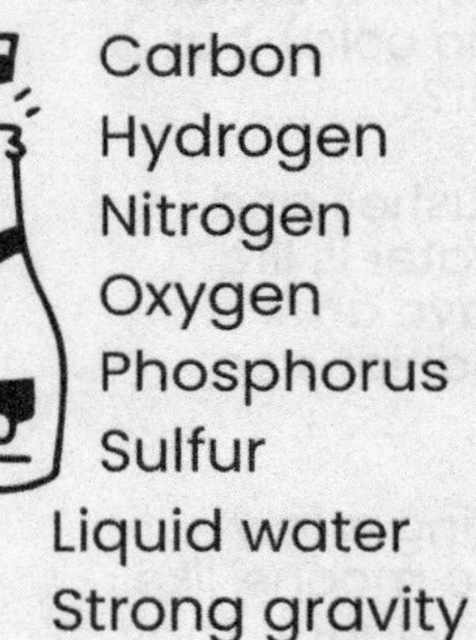

- Carbon
- Hydrogen
- Nitrogen
- Oxygen
- Phosphorus
- Sulfur

Liquid water
Strong gravity

Directions

Mix these on a rocky world that has a magnetic field and atmosphere that orbits a stable star.
Preferably not a red dwarf star if you want longevity, but they are easier to find.

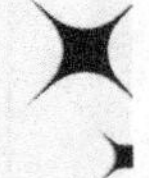

Set to "Goldilocks zone" and let bake for approx. 1 billion years.

Life

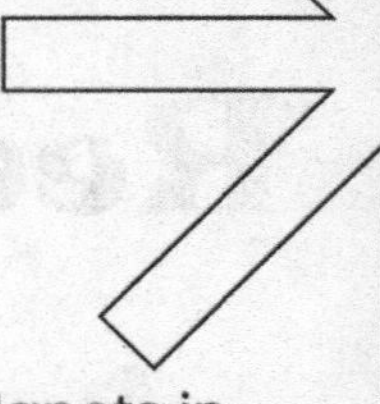

People often ask, "Why do we look for Earth-like planets in the search for life?" I say, "Mofo, do you see life elsewhere in our solar system? If it ain't broke, don't fix it."

The habitable zone is also known as the more fun name the Goldilocks Zone. This is the distance from a star where a planet can have liquid water. It's not too close to the star (not too hot), nor too far from the star (not too cold), but just right for water to exist in liquid form. Get it?

The water molecule is essential because it pushes and pulls nutrients through our bodies. We say "water is life" because all life on Earth needs it. Since we have an abundance of life here, we look for these conditions on other planets and moons to find signs of life.

Space agencies like NASA are keen on exploring a few moons in our solar system for life! Why? Some moons, like Jupiter's moon Europa and Saturn's moon Enceladus, may have a warm liquid ocean underneath their thick ice sheets. Again, water is life. The force of their planet's gravity causes them to squish and stretch. That friction would allow for the subsurface ice to melt into liquid water. If there is life, it could be supported by hydrothermal vents, like how some life at the bottom of our oceans don't get sunlight for energy and have to live off molecules coming from these vents instead.

More than likely, any alien life we find would be as simple as single-celled or extremophiles, as they can survive in more hostile environments.

At the time of writing this, the James Webb Space Telescope detected the presence of carbon dioxide in a region of Europa where material is exchanged between the subsurface ocean and the icy surface. The CO2 is a result of activity by the moon itself and wasn't brought by external sources! As we know, carbon is an essential element for life. This doesn't indicate that life exists, but that life *could* exist. NASA's Europa Clipper mission is set to launch in October 2024. It'll do some flybys and give us more insights into the moon, so let's see!

Edge of the SS

It's still a mystery as to how our planet got its water.

The most accepted explanation is that it was brought here by comets and other small bodies that carried the molecule to Earth during its formation. Earth had the right conditions to keep its water safe from being totally screwed by the Sun. Water depositing would have happened to all the other planets, but they didn't have the right conditions to keep it. For instance, Mars probably had water at some point, but its lack of a good magnetic field made it lose its atmosphere and water to the Sun's wrath.

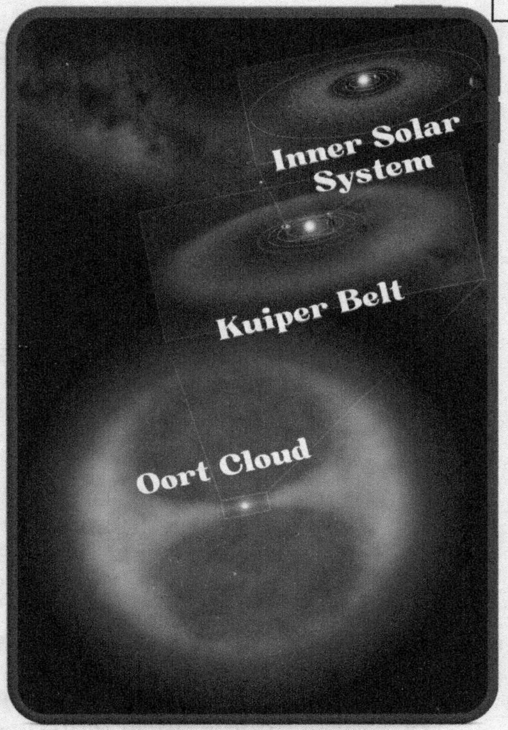

Earlier I mentioned the Oort Cloud. It's insanely massive, remember? It's a giant spherical shell of trillions of icy pieces and space debris. Or, that's what's speculated, as it still needs to be confirmed, but this is where long-period comets must come from. These comets take at least 200 years to orbit the Sun once. Comet Hale-Bopp, anyone? Scientists have yet to observe any object IN the cloud itself. We see them when they come to the inner solar system.

The Oort Cloud is so damn far away that even though the Voyager 1 spacecraft left Earth in 1977 and travels over 1 million miles (1.6 m km) each day, it would take 300 years to reach the Oort Cloud and then about 30,000 years to exit it!

Pluto

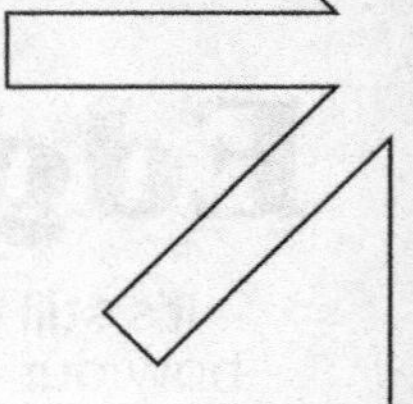

To get to the point, Pluto is not officially considered a planet. It was reclassified to being a dwarf planet or a planetoid in August 2006 by the International Astronomical Union (IAU). According to them, it didn't meet 1/3 points the organization uses to define a full-size planet.

- It is in orbit around the Sun.
- It has sufficient mass to assume hydrostatic equilibrium (a nearly round shape).
- It has "cleared the neighborhood" around its orbit.

Pluto meets the first two points but fails the last. It is not the most gravitationally dominant object in its line of orbit, as it intersects Neptune's orbit and falls into the Kuiper Belt neighborhood.

Other notable dwarf planets include:

- Ceres is located in the asteroid belt.
- Makemake is located in the Kuiper Belt.
- Eris, located in the Kuiper Belt. Its discovery prompted the reclassification system.
- Haumea is located in the Kuiper Belt.

As divisive as this subject is, the reclassification was probably necessary. If Pluto is considered a planet, so should thousands of Kuiper Belt objects (KBOs). Do y'all want to learn the names of thousands of planets in grade school?

Regardless, there is a debate about its reclassification between experts. Dr. Alan Stern, the principal investigator of the New Horizons Mission, the spacecraft that flew by Pluto in July 2015 and gave us the first clear images of it, disagrees with the IAU! He said that "dwarf trees are still trees." He is a planetary scientist. Maybe he'd know better about planets than astronomers? The debate rages on.

New Horizons launched the same year Pluto was reclassified.

Planet IX

No, this is not the conspiracy planet Nibiru. Planet 9 is a hypothetical planet thought to exist deep within the Kuiper Belt by astronomers at Caltech.

- It would be Neptune-sized.
- Its orbit around the Sun would take between 10,000-20,000 years.
- It would have a mass of about ten times the Earth.
- They predict the existence of this object because its gravity could explain the unusual orbits of some dwarf planets and other small icy objects. As in, these objects' paths could be influenced by the gravity of this world.
- It has yet to be discovered officially, but efforts continue.

Mike Brown is one of the astronomers at Caltech leading this search. He's also known as "the man who killed Pluto" because he discovered dwarf planet Eris and others beyond Neptune.

Stars

A Star is Born

By now, you realize how massive a star's influence can be, but let's define a star. I'm sure those QVC hosts could use a refresher, too.

A star is a naturally self-luminous, spheroidal, gaseous celestial body of significant mass that produces energy through nuclear fusion. Basically, it's a giant ball of hot gas that fuses elements in its core.

A star is a delicate balance of gravity pulling stuff in and nuclear fusion in the core pushing pressure out.

Some planets, like Jupiter, are gaseous and round, too. Are they stars? No, dumbass. Jupiter is a planet. It didn't acquire enough mass to ignite nuclear fusion in its core and isn't self-luminous.

They are mostly made of hydrogen, some helium, and a few other elements. The ratios match what makes up the Universe, 73%, 25%, and 2%, respectively. They start by fusing hydrogen in their cores, which lasts most of their lives until they run out of it; at this time, they begin to fuse helium. As the element they fuse gets heavier and heavier, it takes less and less time to do so.

A star's lifespan, death type, and fusion times depend on its mass.

Stars are made in gigantic regions of gas and dust called nebulae, plural for nebula. The most dazzling space pics are of nebulae.

JWST/NASA

Stellar Classification

Oh, be a fine girl, kiss me. Don't worry. I didn't just have a stroke. This sentence forms from the acronym, O, B, A, F, G, K, and M, which is how stars are classified, known as spectral type. Spectral analysis is like taking the fingerprints of a star. You can learn a lot about the star by spreading out the light like a rainbow. The letters are arranged in order of decreasing temperature.

- SC also includes luminosity or brightness classes denoted by Roman numerals, which indicate their luminosity or brightness compared to other stars of the same spectral type, and their evolutionary stage. They are supergiants (I), bright giants (II), giants (III), subgiants (IV), and main sequence stars (V). Stars move off the main sequence into the other categories as they age.
- Temperature subdivisions within each spectral type are denoted by 0-9, with 0 being the hottest.

And let me give a shoutout to Annie Jump Cannon, who created the classification of stars based on temperatures in 1901!

Class	Color	Temps (C)	Size (Suns)	% (stars)
O		>30k	10-100	.00003
B		10-30k	2-20	.13
A		7.5-10k	1.5-2.5	.6
F		6-7.5k	1.2-1.8	3
G		5-6k	1-1.5	7.6
K		3.5-5k	.7-1.2	12.1
M		<3.5k	<.7	76.5

For example, our Sun is a G2V star. It's a main sequence (V) yellow dwarf (G) star that's slightly cooler (2) than a G0V star.

HR Diagram

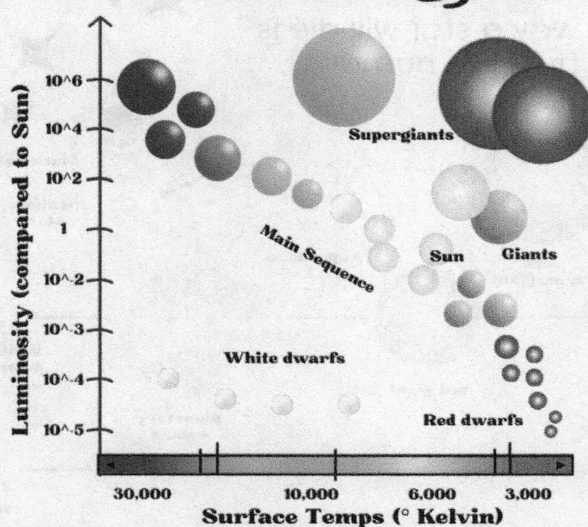

Stars come in all sizes, just like these days in Hollywood. (Don't cancel me.) But now you know that the real stars in space are classified by their surface temperature (x axis) and luminosity (y axis). This chart is the HR Diagram. From left to right the star gets cooler. From top to bottom the star gets dimmer.

Clearly, there's a relationship between a star's temperature and its luminosity (brightness). The bluer the star, the hotter and brighter. The redder the star, the cooler and dimmer. There are always exceptions. Red supergiants can be super bright, too. This is because they are ginormous.

The main sequence is where about 90% of stars in the Universe fall. Their mass can range from 1/10th the mass of the Sun to 200 times its mass. In this stage, they are fusing hydrogen into helium. In about 5 billion years, when the Sun reaches the end of its ten billion-year life, it'll "leave" the main sequence and fall into the (red) giant category.

White dwarfs are leftover cores of stars that have exhausted their nuclear fuel. They primarily comprise electron-degenerate matter, where electrons are packed so closely together that they resist further compression. They're small and dense. Take the mass of the Sun and compress it to the size of Earth! They say that 1 tsp of it weighs as much as three elephants. Their temperature is ~ 180k°F (100k°C), but they'll gradually cool over time. They live for tens or even hundreds of billions of years. This will be the result of our Sun.

This diagram also tells us the lifespan of stars. I always ask my audiences which kind of star would die first, a red dwarf star or a blue star? They're mostly silent, but sometimes they get it right. A blue star would die first because it's burning through its fuel so much more quickly than a cooler, smaller star. Think Hummer vs. a Smart Car.

- Blue stars: live for millions of years.
- Yellow stars (SUN): live for billions of years.
- Red dwarfs: maybe trillions of years. We've never seen one die. They make up ~ 80% of all stars in our galaxy.

A Star is Dead

As mentioned before, the way a star will die is determined by its mass (just like humans).

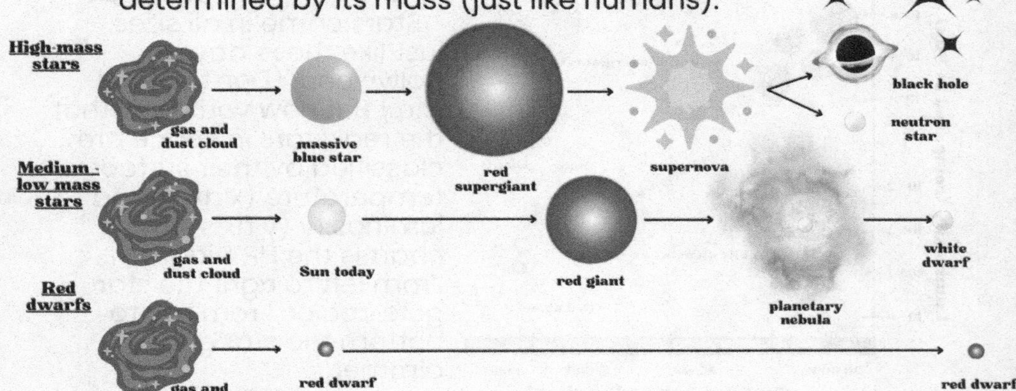

Low-Mass Stars: (< 8 Times the Sun's Mass)

- These kinds of stars, including our Sun, spend most of their lives on the main sequence, fusing hydrogen into helium in their cores.
- As it exhausts the hydrogen fuel, it expands and becomes a red giant. It fuses helium into heavier elements like carbon and oxygen during this phase.
- It ejects its outer layers (a planetary nebula) in the late stages. What remains is a white dwarf.

Intermediate-Mass Stars: (8 - 25 Times the Sun's Mass)

- These follow a similar path to low-mass but more quickly.
- They continue to fuse heavier elements until they reach iron, which cannot support further fusion. The iron core then collapses catastrophically under gravity, leading to a supernova explosion. The core remnants can become neutron stars or black holes on the higher mass end.

High-Mass Stars: (More than 25 Times the Sun's Mass)

- Due to their higher fusion rates, these mofos have much shorter lifespans on the main sequence. They fuse elements up to iron in their cores as well.
- They also end their lives in a supernova explosion. The core collapse is even more violent, resulting in a highly energetic supernova that can outshine an entire galaxy.
- These can end up as black holes.
- **Wolf-Rayet stars** are among the most massive, rare, and bright stars. They lose a ton of mass to their intense stellar winds and may be at the stage before a supernova.

Massive Stars: (Several Hundred Times the Sun's Mass)

- They can end their lives in hypernovae, which are even more powerful than supernovae and can leave behind massive black holes. They're rare.

And **red dwarfs** just chill. After possibly trillions of years, they'll burn through their fuel and end up as white dwarfs.

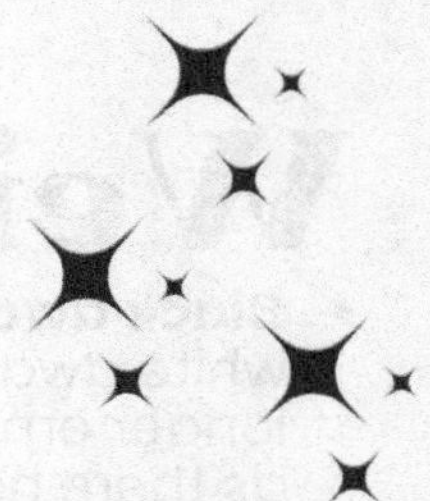

Supernovae

A supernova represents a violent, explosive, yet beautiful end to a star's life cycle. As I mentioned, only stars with a mass > eight times that of the Sun will end this way. Size matters, y'all. Actually, mass matters, y'all.

There are two broad types of supernovae, Type 1 & Type 2.

- **Type Ia** occurs when a star (white dwarf) gathers matter from a companion star, eventually igniting a runaway nuclear reaction. These standard candles are great for helping determine distances in space with their relatively constant brightnesses.
    - **Type 1b and 1c** are stars that lose their outer envelopes of hydrogen and hydrogen and helium, respectively before they collapse.
- **Type II** is when a massive star's core begins to collapse, and the bounce back of the implosion expels the outer layers of lighter elements. This can briefly outshine an entire galaxy. These are the ones people typically think of when they hear *supernova*.

On average, a supernova occurs once every 50 years in our galaxy.

Weird Stars

- **Black dwarfs**: after maybe trillions of years, a white dwarf will cool into a black dwarf. These no longer emit heat or radiation. We've never seen one, as there hasn't been enough time for this process to occur, and it would be damn difficult to "see" one anyway.

- **Brown dwarfs**: the losers who couldn't get enough mass to ignite nuclear fusion. They could be the size of Jupiter or larger but with a mass 13-80 times greater. They lack nuclear fusion, so they emit almost no light.

- **Betelgeuse**: not a weird star, but it's notable. This red supergiant makes up the left shoulder of the Orion constellation. It keeps giving signs that it will go supernova, which is something we'd be able to see with the naked eye! Of course, we'd want that to have happened about 750 years ago so that light would reach us in our lifetime.

- **Neutron stars**: the result of a very massive star exploding. They're only about 12 miles (19 km) in diameter but much more dense than the Sun. They are so dense that one teaspoon weighs as much as Mt. Everest! (I love these comparisons.) The gravity is so strong that the protons and electrons squish together into neutrons.

- **Magnetars**: a rarer type of neutron star with extreme magnetic fields. The fields are so powerful that being within 600 miles (965 km) of one will destroy you down to the atomic level by distorting the electron clouds. Basically, you'll disintegrate.

- **Pulsars**: the "lighthouses" of the Universe, as they are rapidly rotating neutron stars that shoot out two jets of particles. They can spin at least 700 times in 1 second!

- **Erendal**: the most distant star ever observed. Its light has taken 12.9 billion years to reach us.

- **Methuselah**: the oldest star at maybe ~ 13.65 billion years old but with a large margin of error in age.

- Most stars are binary. (I'm not even going to go there.) They come in pairs and revolve around each other, or the smaller one revolves around the larger one.

- Black holes are stellar corpses, so we'll throw them here too. More on them later.

Stardust

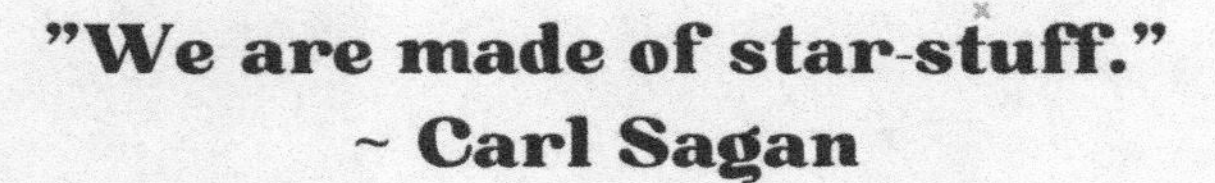

"We are made of star-stuff."
~ Carl Sagan

It may be cheesy as hell, but we are made of stardust. It's not just us, but almost everything we give value to wouldn't exist without stars and their deaths.

Stars are the only things powerful enough in the Universe to fuse elements. We get things like carbon, silver, gold, iron, silicon, etc. Many of these elements make up our valuables and possessions, and many make up our bodies.

From inside stars, we get all elements from helium to iron. From supernovae, we get all of the elements from cobalt and beyond. Remember, you read the periodic table from left to right.
The elements that makeup 99% of humans are hydrogen, oxygen, nitrogen, carbon, calcium, and phosphorus. Sulfur, potassium, sodium, chlorine, and magnesium make up the remaining < 1%.

I'm basically spelling it out for you guys. The elements that create humans overlap between the elements made in stars and those made from supernovae.

So, you are made of stardust.

Elements made inside stars end. **Elements made from supernovae begin.**

H																	He
Li	Be											B	C	N	O	F	Ne
Na	Mg											Al	Si	P	S	Cl	Ar
K	Ca	Sc	Ti	V	Cr	Mn	Fe	Co	Ni	Cu	Zn	Ga	Ge	As	Se	Br	Kr
Rb	Sr	Y	Zr	Nb	Mo	Tc	Ru	Rh	Pd	Ag	Cd	In	Sn	Sb	Te	I	Xe
Cs	Ba	La	Hf	Ta	W	Re	Os	Ir	Pt	Au	Hg	Tl	Pb	Bi	Po	At	Rn
Fr	Ra	Ac	Rf	Db	Sg	Bh	Hs	Mt	Ds	Rg	Cn	Nh	Fl	Mc	Lv	Ts	Og
			Ce	Pr	Nd	Pm	Sm	Eu	Gd	Tb	Dy	Ho	Er	Tm	Yb	Lu	
			Th	Pa	U	Np	Pu	Am	Cm	Bk	Cf	Es	Fm	Md	No	Lr	

Galaxies

The Milky Way

At least once in your life, go to a dark sky area and see the band of the Milky Way at night. Dark skies are disappearing, so go now!

Our Milky Way galaxy is a spiral galaxy. It's thought to be between 100k-150k light years from end to end. Remember, a light year ~ 5.8 trillion miles (9.3 t km). Obviously, we don't know its exact size because we're in it. Since it's insanely massive, we have yet to go anywhere near being far away from it to turn around and take a picture. Imagine a T-rex trying to take a selfie.

Our solar system is in one of the spiral's arms called the Orion Arm. It's the 4th arm from the center. Are we in the boondocks? Maybe. Perhaps that's why aliens don't want to visit... among other reasons. We are about 25k light years from the galactic center and its supermassive black hole known as Sagittarius A Star. We got an actual image of that black hole in 2022!

The Milky Way is estimated to have <u>at least</u> 100 billion stars and formed about 13.6 billion years ago, soon after the Big Bang. The name comes from Greek mythology. The very summarized story says that Heracles (Hercules is the Roman name) bit Hera's tit and milk sprayed out across the heavens. Hence, the band of the *Milky* Way.

In 5 billion years, the Milky Way will collide violently with our neighbor, Andromeda, forming an elliptical galaxy. **#Milkomeda**

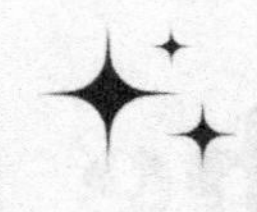

Hubble's Law

In the 1920s, Edwin Hubble figured out that there are galaxies outside of our Milky Way. Until then, these fuzzy patches of light were broadly labeled "nebulae." Scientists speculated that maybe there were other "islands of stars" outside of the Milky Way. Still, it wasn't until Hubble used Mount Wilson Observatory's 100-inch Hooker Telescope, the largest telescope at that time, that we got definitive proof that there were other galaxies beyond our own!

He looked at Andromeda, our galactic neighbor at 2.5 million light years away, and found that it too contained its own individual stars that were well beyond the stars in our galaxy. Conclusion? This "nebula" was actually its own galaxy as were all the other "nebulae" out there!

Up until these discoveries, it was believed that we lived in a static Universe. As in, it was more or less the same throughout eternity. However, Hubble discovered that our Universe was expanding uniformly in all directions! These newly discovered galaxies were moving away from us, and more distant galaxies were moving away even faster. This is now known as Hubble's Law. A galaxy's velocity is directly proportional to its distance.

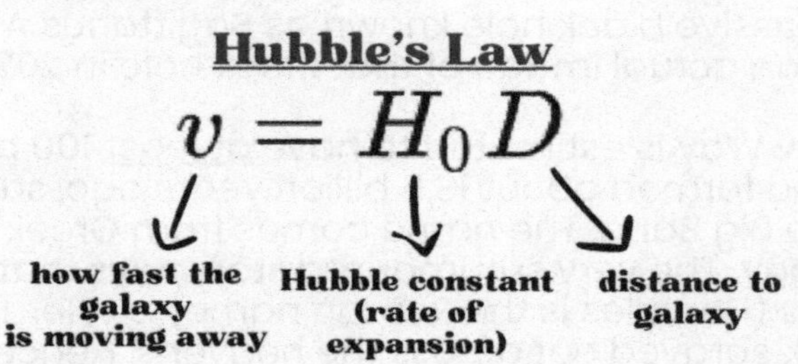

The Hubble constant, or expansion rate, has changed since Hubble's time. It's damn tricky to get an exact rate, but it generally falls in the range of approximately 67 to 74 kilometers/second/ megaparsec. (A megaparsec ~ 3.26 million light years.) Again, the further away a galaxy is, the faster it moves away from us. This is calculated by the galaxy's redshift, which I get into later.

Galaxy Types

A galaxy is a collection of millions to trillions of stars, gas, dust, and maybe dark matter (more later) held together by gravity. Most house a supermassive black hole at the center. It is said that there is at least one planet to every star, which means that galaxies can house billions or trillions of planets. (We just CAN'T be alone!) Estimates put the number of galaxies in the observable Universe between 200 billion to 2 trillion. It's a large range, but you try to do better.

Types include:

- **Spiral galaxies**: a primarily flat rotating disk with spiral arms that curve out from a central bulge.
- **Elliptical galaxies**: thought to result from mergers of spiral galaxies, these kinds are older as they house little gas and dust to form new stars. They have little order, and they can be round and oval.
- **Lenticular galaxies**: also thought to be older, these look like a mix of spirals and ellipticals as they are spiral-disk in shape but with no arms. They lack the gas and dust to make new stars.
- **Irregular galaxies**: the weirdos of the Universe as they range in shape and can have a few hundred million stars to tens of billions. These are thought to be created when two galaxies do a driveby, and one gets messed up by the other's gravity. They typically have a mix of old and young stars.
- **Quasars**: the most luminous type of active galactic nucleus (AGN) that shoots out powerful jets and radiates thousands of times the energy that our Milky Way does. These are more than likely powered by an active supermassive black hole.
    - **Blazars** have the coolest name and are quasars whose jets point directly at Earth.

~ 10% of galaxies are active, meaning the supermassive BH at the center is "feeding" on its surroundings.

Galaxies are missing visible matter (baryonic matter)! A certain amount of matter is calculated to exist in our Universe. Since galaxies are the largest collections of matter, they should house most of it. However, they only contain half of what they should. Evidence shows that this matter has been dispelled into the space between galaxies.

Recent findings by the James Webb Space Telescope have called into question *galactic formation* (not the age of the Universe) since it has found well-formed galaxies that existed much earlier than we ever thought possible.

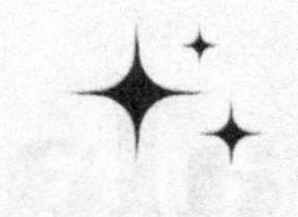

Measuring Distances

Not all humans are intelligent. Actually, most aren't. However, the few who are smart use math to calculate distances in space. Remember, astronomy is approximation.

- **Parallax**: for objects within our solar system and stars in our galaxy.
  - The apparent shift in the position of a nearby object against a distant background as the Earth orbits the Sun. Try it! Hold out your thumb and close one eye at a time. Notice how it shifts compared to the background.

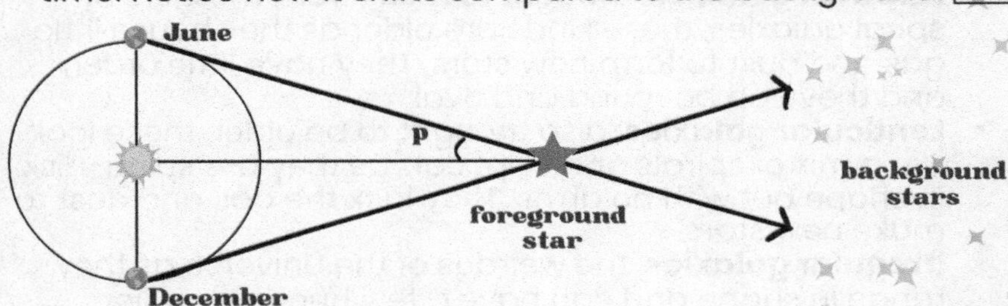

  - The line of sight differs from June to December. Half of the angle this creates is the parallax.
  - Astronomers use d=1/p and trigonometry to solve.
- **Standard Candles**: objects with known luminosities. You can calculate the distance by measuring the apparent brightness of these objects.
  - Examples include Type 1a supernovae (previously mentioned) and Cepheid variable stars.
  - Cepheids are a class of pulsating variable stars. They brighten and dim over the same amount of time.
  - Henrietta Swan Leavitt was the first to notice a <u>relationship between brightness and distance</u> in 1908, a huge discovery that allowed us to calculate distances!
  - If you know how bright the star is, you can solve for the distance.

$$d = \sqrt{\frac{L}{4\pi B}}$$

L = luminosity, true amount of energy being given off
B = brightness, how bright it appears to us

- **Redshift**: the stretching of light waves towards red as the object moves away from you.
- **Gaia Spacecraft**: giving us the most accurate map of 2 billion objects in the Milky Way!

Exoplanets

Exoplanet Overview

An exoplanet is any planet outside of our solar system. Since the early 90s, we've discovered over 5,000!

Types of exos so far include:

- Gas giants: some even larger than Jupiter.
- Super Earths: they wear a red cape. JK. They are rocky worlds larger than Earth.
- Neptune-like: pretty self-explanatory.
    - Hycean planets are hot, covered with water, and have Hydrogen-rich atmospheres.
- Terrestrial: basic rocky world.

There are ranges of sizes within all of these types.

The first exoplanet to gain fame was 51 Pegasi b in 1995. It's a hot Jupiter orbiting a Sun-like star about 50 light years away.

Hot Jupiters were considered the most abundant exoplanet type, but some say Neptune-like and even mini-Neptunes are more common.

The closest exoplanet to us is Proxima Centauri b. It orbits our nearest star neighbor at only 4 light years away and is a super Earth. It is in the habitable zone but is tidally locked so the same side always faces the star. More than likely it's uninhabitable, but more research is needed.

Although the info varies, some say the furthest exoplanet discovered is SWEEPS-11, a hot, large Jupiter about 27,000 light years away.

Most planets are believed to be rogue, meaning they wander the galaxy alone without orbiting any star. They would have formed around a star but got flung out for whatever reason. So now, they're pretty much homeless. It's estimated that there are six times as many rogues as orbiting planets. (Just like downtown LA.) It's improbable that they would have any life since there's no energy source (star). I have yet to find an exact answer, but we've discovered at least 70 rogue planets. This is tough to do since there's no light for them to reflect!

Finding Exos

These are the five methods astronomers use to find exoplanets... explained in a basic a** way. They are listed from most to least discoveries made.

1) **Transit Method**: (4,132 exoplanets discovered so far)
When an exoplanet orbits in front of its star, the light output from that star to us dips a little. Based on the measured dip, one can infer an exoplanet.

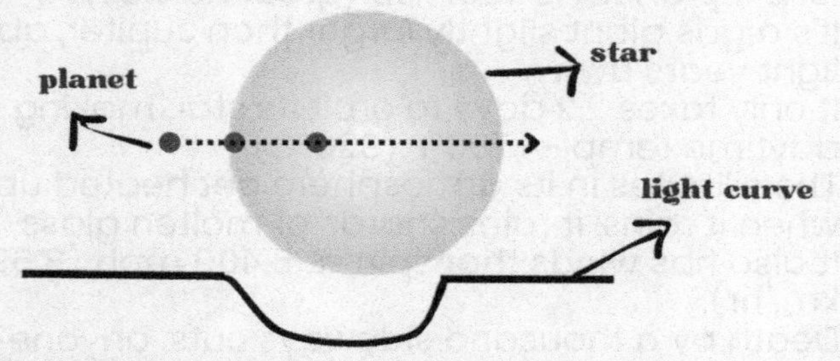

2) **Radial Velocity**: (1,068 exos so far)
Even planets exert some gravity on their host stars, albeit small. Scientists can measure the star's "wobble" and infer an exoplanet.

3) **Gravitational Microlensing**: (204 exos so far)
This one's tough to imagine, so sucks to suck if you don't get it. Light can bend around objects in space. Einstein was brilliant because he knew that. When a star and its planet pass between us and a more distant star, the light from the distant star distorts from our POV. Catch that distortion and you can infer an exoplanet in the closer system.

4) **Direct Imaging**: (69 exos so far)
It's self-explanatory as this is where astronomers block out the glaring light of a star and snap a pic of the exoplanet(s) that orbit it.

5) **Astrometry**: (3 exos so far)
Think of the wobble from the radial velocity method but on a wider scale. This method tracks a star's wobble in relation to other stars around it. It's difficult to do because it requires such highly precise optics that our atmosphere screws up.

Telescopes used for the exoplanet hunt by NASA: Hubble, Spitzer, Kepler, Chandra, TESS, and JWST.

Weird Exos

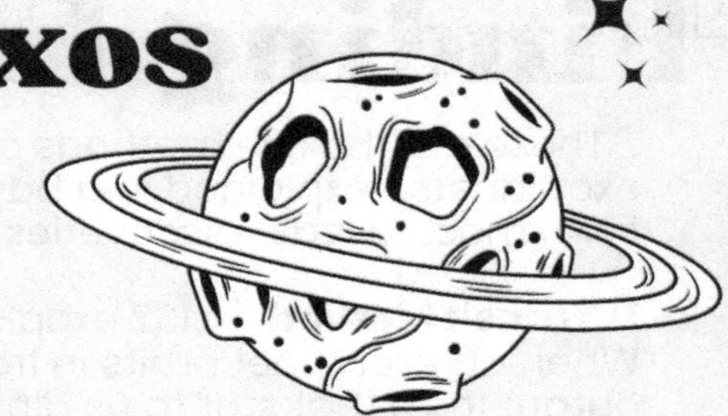

This should be fun.

- Worst exoplanet: HD 189773b (great names)
  - It's a gas giant slightly larger than Jupiter, about 63 light years away.
  - It only takes 2.2 days to orbit its star, making its daytime temp ~ 1,700°F (926°C).
  - The silicates in its atmosphere get heated up, so when it rains, it rains shards of molten glass.
  - It also has winds that spin at 5,400 mph (8,690 km/hr).
  - Death by a thousand sideways cuts, anyone?

- Fluffiest exoplanet: TOI-3757 b
  - It's also a Jupiter-type exoplanet about 580 light years away.
  - Its atmosphere has the density of a marshmallow.

- Luckiest exoplanet: Halla (8 UMi b)
  - Once again, it's another Jupiter-type exoplanet about 520 light years away.
  - It was discovered by Korean astronomers in 2015.
  - Its star was in its death process as it was swelling out to become a red giant so it should have gotten eaten up, but somehow it escaped certain doom.

- Bougiest exoplanet: 55 Cancri e
  - This one's a super Earth about 40 light years away.
  - It's composed mainly of carbon (in the form of diamond and graphite), iron, silicon carbide (when carbon is combined with any metallic or semimetallic element), and potentially silicates.
  - Around 1/3 of its mass is diamond! (Good thing it's too far to be mined.)

*Disclaimer - this composition may no longer be correct, but "diamond" exos exist.

Telescopes

EM Spectrum

Since we will be talking about telescopes, this is the perfect place to discuss the electromagnetic spectrum.

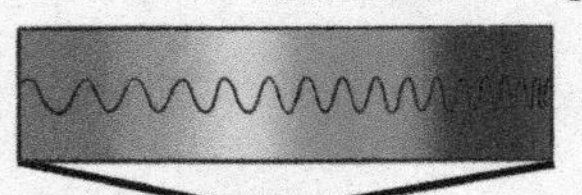

Radio Microwave Infrared Visible Ultraviolet X-Ray Gamma Ray

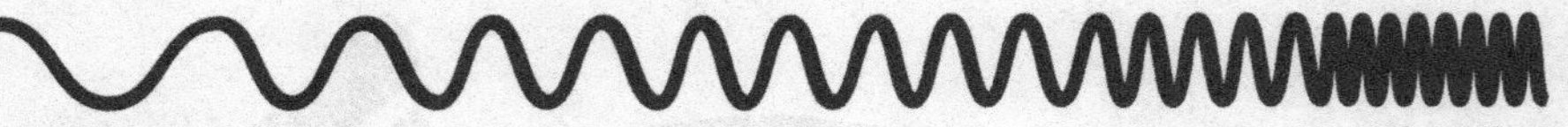

wavelength = hundreds of miles (km)

wavelength = smaller than atomic nuclei

I also call it the light spectrum. From left to right, or radio waves to gamma rays, the waves go from long and low frequency to short and high frequency. In other words, they get more dangerous. They all travel at the speed of light ~ 186,000 miles (300,000 km) per second, or 7.5 times around the Earth in 1 second!

FUN FACT

The sky is blue because blue light has shorter wavelengths than red light; therefore, blue light gets scattered more easily by the particles in the atmosphere. This is also why sunsets are reddish. The blue light has been scattered away.

Humans suck, so we only see a small part of the whole spectrum known as visible light. Our eyes don't see most of all light. Some animals, like bats, rodents, and reindeer, and probably many others, see in ultraviolet. Others see in infrared. Both would be cool!

That's why there are various types of telescopes in space.
JWST (James Webb Space Telescope) sees in infrared.
Hubble sees primarily in visible light but in some infrared and ultraviolet.
VLA (Very Large Array) is a radio telescope in New Mexico.

Wait, are radio waves and radios that play music related? Radios receive electromagnetic radio waves, extract the audio information encoded in those waves, and convert that information into sound waves we can hear. Radio waves are the best option for this due to their long wavelengths, low energy, and easy penetration of Earth's atmosphere, among other properties.

Our atmosphere filters out many waves, so having various telescopes in space is best. #canceltheatmosphere?

Spectroscopy

I touched on this briefly with stars, but let's dive deeper into <u>spectroscopy</u>, which is the study of the absorption and emission of radiation (light) by matter. You have to split the electromagnetic radiation into the wavelengths that make it up, like how a prism breaks visible light into a rainbow.

This is how scientists determine what elements exist in or around whatever object you're looking at. It also tells us the gases in planets' atmospheres, the speed and temperature of gases, the rotation speed and composition of stars, and more!

The key components that are needed are electromagnetic radiation and electrons in atoms.

- Electron(s) around atoms orbit in very specific levels.
- Electrons can only transition to another level if they absorb or emit light at <u>specific</u> wavelengths.
- They can't go in between these levels.
    - For instance, to jump to the 2nd step from the ground, I'd need to eat 1 teaspoon of cake (I love cake). To jump to the 5th step from the 2nd, I'd need to eat 3 tsps of cake. To return to ground level from the 5th step, I'd need to throw up 4 tsps of cake, not 4.5, not 3.75, precisely 4! Each level requires an exact amount of cake.
- <u>Photons</u> are particles of light that carry a specific amount of energy corresponding to their wavelength. Violet light (shorter wavelength) has more energy than red light (longer wavelength) and would make an electron jump to a higher level than red light. (MORE CAKE, MORE JUMP!) Electrons absorb or emit these specific photons.
- Each element on the periodic table has unique energy levels for its electrons.
- The wavelengths the electrons absorb or emit show us the unique "fingerprints" of each element.

Emission vs. Absorption Spectra

Spectra of the awesome "Kalpanium" element.

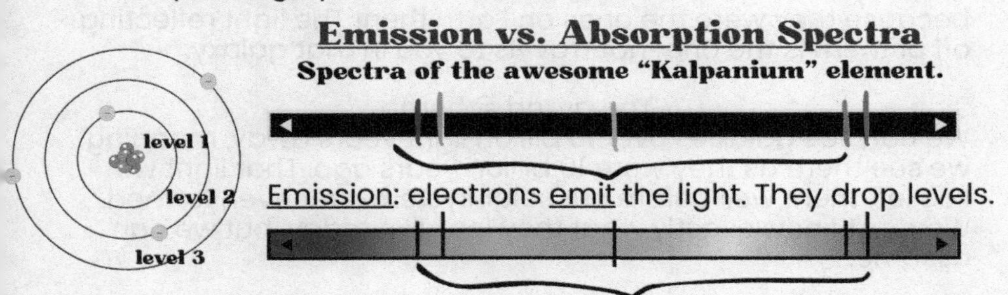

<u>Emission</u>: electrons <u>emit</u> the light. They drop levels.

<u>Absorption</u>: electrons <u>absorb</u> the light. They jump levels.

This is how we'd know that the Kalpanium element is present in whatever object we were looking at. These spectra are its "fingerprints." They would look different for every element!

Light, Fast AF

One of my favorite things to talk about is the speed of light, because not only is it the fastest thing in the Universe, traveling at 186,000 miles (300k km) per second, but it can also take you back in time... visually.

Even though light is fast af, space is still stupidly huge! Therefore, even light takes time to traverse such great distances. That's why we use light years to measure distances in the cosmos, as we might as well use the fastest speed possible. When we look at any object in space, we look at it how it was in the past. The light that left it "x" amount of time ago is the one that's reaching your eyes now.

Let's start locally.

Our Sun is about 93 million miles (149 m km) away, so it takes its light about 8 and 1/3 minutes to reach our eyes. This means that whenever we see it, we look at it as it was 8 1/3 minutes ago. That sunset? It's already set!

Zooming out.

Markarian's Chain is a cluster of galaxies that vary in distance, but let's pick one that's about 66 million light years away. Remember, a light year is 5.8 t miles (9.3 t km). When we look at this galaxy, we see it as it was 66 million years ago because that light traveled for that length of time to get to us. If you lived in that galaxy and had a powerful enough telescope to see Earth and somehow its surface; what would you see? If you paid attention to "66 million years ago", you'd know that you'd see dinosaurs because they were the ones on Earth then! The light reflecting off of them is the one that travels to you in that galaxy.

The grand scheme.

We can see galaxies over 10 billion light years away, meaning we see them as they were 10 billion years ago. That light we see left them well before our solar system ever even formed. We don't know exactly what they look like today, but we can assume.

The James Webb Space Telescope recently discovered a galaxy named JADES-GS-z13-0. Measurements say it formed 320 million years after the Big Bang, making it the oldest galaxy found so far. Therefore, we see it as it was 13.5 billion years ago when the Universe was only 2.3% of its current age. So crazy!

Dark Skies

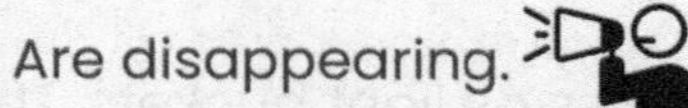

Are disappearing.

Artificial lights, mostly street lights, are causing dark skies to disappear. These lights interact with the atmosphere and cause night glow.

At the rate we're going, all the stars in the night sky could disappear in 20 years.

The Milky Way is no longer visible to 1/3 of humanity.

This not only affects our ability to gaze upon the wonders of the Universe, but it's messing with animals as well.
*Also, non-human animals > human animals. Comet me, bro.

Many animals depend on moonlight to guide them, but artificial light can trick them. This disrupts migration patterns, resulting in millions of deaths every year!

The blue light from LEDs also affects our sleep cycles. Some studies say that the lack of red light, which can penetrate deeper into our bodies, can be linked to increased obesity.

Solutions: point lights down, put fewer around, and only turn them on when needed.

A list of only some affected animals:

- Sea Turtles
- Frogs and Toads
- Humming Birds
- Zebrafish
- Sweat Bees
- Monarch Butterflies
- Atlantic Salmon
- Zooplankton
- Bats
- Owls
- Mice
- Fireflies

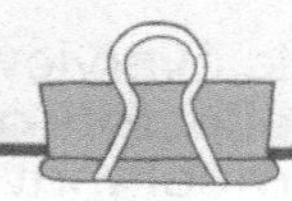

Wouldn't it be great if we had one night a year when all lights had to be turned off?

Types of Teles

Think of telescopes as light buckets. The larger the lens or mirror, the more light you would "catch." The ocean catches more rain than a lake simply because it's larger. Telescopes catch photons (light particles) in the same way.

Terms:

- Chromatic aberration: fringing of colors as light from different wavelengths split and arrive at different angles.
- Aperture: diameter of the light collecting region, the lens or mirror.

For home use, there are two basic types.

- **Refractors**: use lenses.
  - Longer tubes allow for the light to focus.
  - Better for deep-space objects.
  - Must account for chromatic aberration.
  - Smaller aperture for affordability.
  - Low maintenance.

- **Reflectors**: use mirrors.
  - Larger aperture that doesn't require a longer tube.
  - Better for larger and brighter objects (moon, planets).
  - No chromatic aberration.
  - Bulkier.
  - Requires more maintenance.

This is a basic overview, but pros have told me that Dobsonian reflectors are the way to go for beginners. Again, remember that it depends on your budget and other parameters! Do a lot of research before you commit.

The OGs

NASA's Great Observatories Program is a group of 4 space-based observatories that look at the Universe in different wavelengths (light). They paved the way.

- Of course, we have to start with the **Hubble Space Tele**:
    - Launched April 24th, 1990.
    - Orbits 340 miles (547 km) above Earth's surface.
    - Looks primarily in visible wavelengths.
    - Found more moons around Pluto, peered back billions of years in time, studied our solar system, discovered supermassive BHs at the center of galaxies.

- **Spitzer Space Telescope**:
    - Launched August 25th, 2003.
    - Has an "Earth-trailing" orbit.
    - Looked primarily in infrared wavelengths.
    - Found the largest ring around Saturn, 7 exoplanets in the TRAPPIST-1 system, an exoplanet at 13,000 light years away, studied most distant known galaxies.
    - Ended on January 30th, 2020.

- **Compton Gamma Ray Observatory**:
    - Launched April 5th, 1991.
    - Orbited at 270 miles (434 km) above Earth's surface.
    - Looked primarily in gamma rays (duh).
    - Discovered blazars, gamma rays in Earth's thunderstorms, the most distant gamma-ray bursts, antimatter from the center of the Milky Way.
    - Deorbited on June 4th, 2000.

- **Chandra X-ray Observatory**:
    - Launched July 23rd, 1999.
    - Orbits 65k miles (104k km) above Earth's surface.
    - Looks primarily in X-rays (duh).
    - Studied the structure of supernovae, habitability of environments around stars, found X-rays from our supermassive black hole.

You'll see that many teles use infrared because:

- You can see further back in time as light from the early Universe has been stretched to infrared lengths (more on this later).
- It can penetrate through clouds of gas and dust so that we can see the stars and planets forming in these clouds.
- It allows scientists to study cooler, older objects.

Next Gen

Here are some new baddies.

- We have to start with the **James Webb Space Tele**:
  - Launched December 25th, 2021.
  - Hexagonal mirror is 21.3 ft in diameter.
  - Orbits at L2*, 1 m miles (1.6 m km) away from Earth.
  - Looks primarily in infrared.
  - Has seen galaxies as far back as 325 million years after the Big Bang, the oldest black hole, potential of molecules produced by life on an exoplanet, carbon on Europa, and MUCH more to come.

- **Nancy Grace Roman Space Telescope**:
  - Scheduled to launch in May 2027.
  - Panoramic field view 100x larger than Hubble's.
  - Will look in visible and near-infrared.
  - Will delve into dark energy, exoplanet discoveries, and infrared astrophysics.

- **SPHEREx**:
  - Scheduled to launch by April 2025.
  - Will take a map of the entire sky every six months.
  - Will look primarily in infrared.
  - Half the size of Webb's mirror.
  - Will probe what happened 1 second after the Big Bang, how galaxies form and evolve, and search for molecules necessary for life.

- **Euclid** (ESA):
  - Launched July 1st, 2023.
  - Orbits at L2, 1 m miles (1.6 m km) away from Earth.
  - Looks primarily in visible and near-infrared.
  - Will make a 3D map of the Universe by observing billions of galaxies and will delve into the natures of dark matter and dark energy (which I will delve into later).

*L2 is a <u>Lagrangian</u> point, where gravitational forces and the orbital motion of a body balance each other. This balance in forces allows the spacecraft to hover. Also, I focus a lot on NASA stuff because I live in 'Merica, but agencies worldwide are doing amazing things and, quite often, doing things together.

Ground Teles

- Keck Observatory in Maunakea, Hawaii, consists of two telescopes that are 32.8ft across. They look in optical and infrared and can see further away than Hubble. It studies supernovae, dark matter, exoplanets, and distant galaxies, to name a few.

- Atacama Large Millimeter Array (ALMA) in the Atacama Desert, Chile consists of 66 radio telescopes, the majority being 39.4ft across. One of its most significant discoveries was finding oxygen 13.28 billion light years away.

- South African Large Telescope (SALT) in Karoo, South Africa is a 32ft optical telescope. Among its many discoveries was the discovery of the first white dwarf pulsar.

- The Thirty Meter Telescope in Maunakea, Hawaii is a wonderful collaboration between the US, India, China, and Japan, and is currently under development. Obviously, its primary mirror is 30m (98ft) and consists of 492 hexagonal panels. It will study black holes at the center of our galaxy and others.

- The Extremely Large Telescope (don't you love these names?) is also under development and set to be completed by 2027 by the European Southern Observatory (ESO). As its name suggests, its mirror is large at 128ft. Therefore, it can collect 100,000,000 times more light than the human eye. Designed to look in optical and infrared, its objective is to discover Earth-like exoplanets and life beyond our solar system.
    - The ESO is an intergovernmental research organization that is comprised of 16 European countries. This is a different organization than the ESA or European Space Agency.

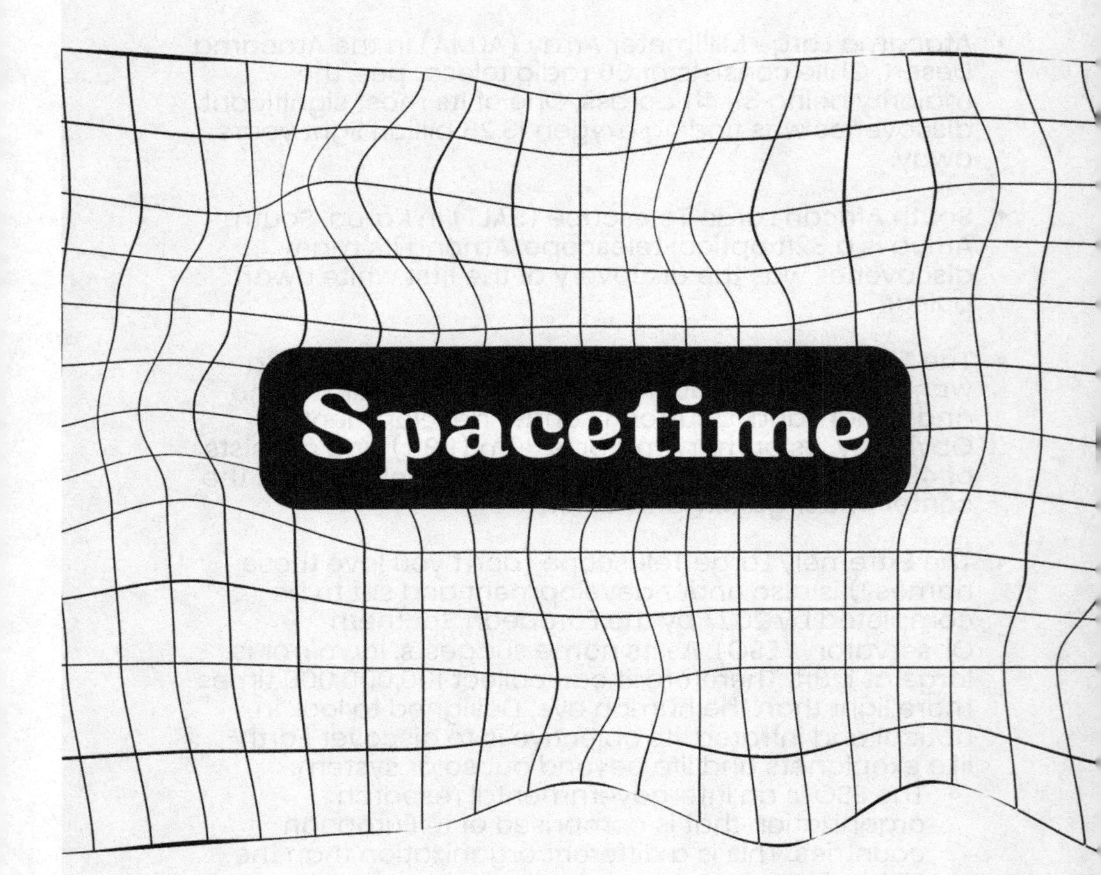
Spacetime

Cosmological Constant

When you think about it, it wasn't that long ago when Hubble discovered that the Universe was expanding. People didn't even know that there were galaxies outside our own until about 100 years ago (in the Western world). The Big Bang theory came about in 1927, but getting to that point (pun intended) about the true nature of our cosmos wasn't easy.

The cosmological constant was first introduced by Albert Einstein in 1917. Denoted by Λ (Greek letter Lambda), it is a fundamental constant in describing the large-scale structure of the cosmos and its evolution. In physics, a constant is a physical quantity that is believed to be universal in nature. Its value doesn't change. Einstein's field equations didn't quite align with the belief at that time that the Universe was static and unchanging, so he added the constant to maintain the static view. Basically, although the math tried to tell him that the Universe *wasn't* static, he added a constant to his equations to make it so.

However, when it was later discovered that the Universe *was* expanding, Einstein removed the constant since his equations actually did align with a changing Universe. He called the initial addition of the CC his "biggest blunder." So dramatic.

Then, in the late 1990s, it was discovered that the Universe was not *just* expanding but expanding faster with time. So, the constant was added back in! Today, the cosmological constant represents the mysterious force of dark energy that's causing the accelerated expansion of space.

I mentioned the Hubble constant earlier, but it's not to be confused with the cosmological constant. They are related, but the HC is the current expansion rate while the CC is the gravitationally repulsive force (dark energy) causing the accelerated expansion. CC would make HC increase over time. Again, HC is a rate, while CC is the reason for the rate.

The cosmological constant fits into how accurately Einstein's field equations describe the nature of matter and gravity and, ultimately, the structure of our entire cosmos.

We'll delve into relativity theory later.

The Big Bang

In a nutshell, this theory states that the entire Universe began as an infinitely hot and dense point that expanded with time. It was not an explosion but a rapid expansion of **spacetime** - 3 dimensions of space (x, y, z) and 1 dimension of time fused. Imagine spacetime as a malleable fabric that permeates the entire Universe.

The insane timeline goes something like this:

- **The Big Bang**: time = 0.
- **Planck Epoch**: time = 10^-44 seconds, when the four fundamental forces were unified (more on these later).
- **Inflation Epoch**: Time = 10^-36, the Universe expanded exponentially.
- **Quark-Gluon Plasma Epoch**: time = 10^-6 seconds, quarks and gluons (the building blocks of protons and neutrons) freely existed and not within particles.
- **Nucleosynthesis**: time = 3 minutes, the Universe was cool enough to form light elements (H, He).
- **Photon Epoch**: time = up to about 380,000 years; the Universe was too hot, so free electrons scattered light. This opaque plasma is a barrier to "seeing" before this.
- **Recombination**: time = after 380,000 years, the Universe cooled enough for free electrons to recombine with nuclei, which freed light. We detect this today as the cosmic microwave background. (More on this next).
- **Dark Ages:** time = after 380,000 - a few hundred million years, the Universe was dark as nothing formed.
- **Reionization**: time = a few hundred million years; the very first stars and galaxies were able to form and emit intense ultraviolet radiation, which reionizes H atoms.
- Gravity continues to draw matter together to form more stars, galaxies, and galaxy clusters while the Universe expands. We continue on this path today.

Science has a solid understanding from the nucleosynthesis era on. The fact that we have any knowledge of these incomprehensible timeframes is mindblowing to me. It gets harder to detail closer to the Big Bang because our Universe was so different at those times, but physics shows us that this is our best model.

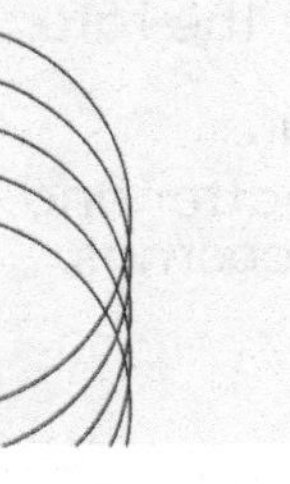

BB Evidence

Contrary to what certain public figures might have recently said, the scientific community still thinks the Universe is about 13.7 billion years old. There is plenty of evidence to support this theory.

- **Redshift of galaxies**: this cosmological redshift shows us that the Universe is expanding. As galaxies move away from us, their light stretches and thus appears redder. If you go in reverse of the Hubble constant (expansion rate of the Universe), eventually everything will come together 13.7 billion years ago, known as Hubble time. More on redshift next.
- **Light elements**: the amount of hydrogen and helium in old stars and galaxies is consistent with an early hot and dense Universe.
- **CMB**: the cosmic microwave background is radiation left over from the Big Bang. The early Universe was hot as hell, so the CMB is leftover heat from around 380,000 years after the BB. This is the farthest back in time we can see! Much time has passed since then, so it is only detectable in microwave wavelengths and is 2.725° Kelvin above absolute zero. It's also a large part of the static on analog TVs!
- **ΛCDM**: is not "evidence" but a parameterization (convenient parameters) of the standard model of the Big Bang. Λ (Lambda) represents dark energy and CDM is cold dark matter. You'll learn more about these later.

These are the conceptual terms that have plenty of math to support them.

The Universe may be 13.7 billion years old, but we're just emerging out of the cosmic womb! Some timelines "end" the Universe at a few thousand trillion trillion trillion trillion trillion trillion trillion trillion years from now. (Way too many zeros.)

CMB

NASA

Redshift

As objects move away from us, the light they emit gets longer and, therefore, redder, and I mean that literally. In the EM spectrum, redder light has longer wavelengths than blue and violet light. When something moves away from you, its light gets redder. This is known as the Doppler effect. We're used to this idea when it comes to sound. Listen to an ambulance drive by next time. It sounds different when it comes closer than when it drives away. It's the same with light. As a galaxy moves away, its light gets more red. You can figure out the distance to an object based on its redshift, "z." The opposite happens too. When a galaxy comes towards you, it's light gets more blue.

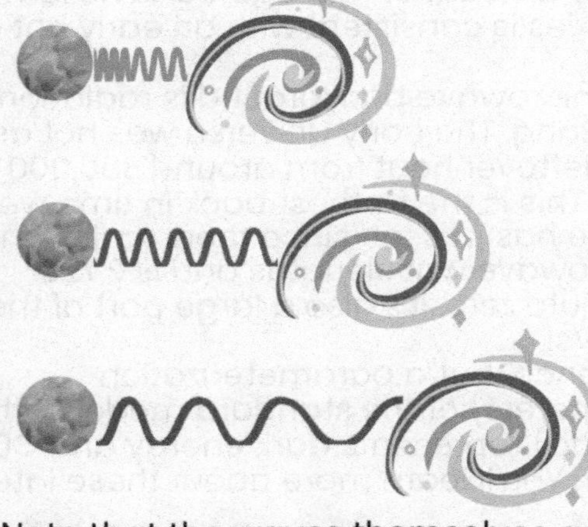

$$z = \frac{\lambda_{obs} - \lambda_{rest}}{\lambda_{rest}}$$

z = **redshift**
λ_{obs} = **observed wavelength**
λ_{rest} = **rest wavelength**

Note that the waves themselves get stretched out. The higher the "z," the farther an object is. The galaxy JADES-GS-z13-0 that I mentioned is the oldest and farthest galaxy ever found. It has a z=13, which is a very high redshift number! The previous record holder was GN-z11, putting it slightly closer. They formed about 13.5 and 13.4 billion years ago, respectively. However, their current redshift distance is 33.6 and 31.96 billion light years away! Since the Big Bang, space has expanded to a diameter of 93 billion light years! What's causing it? That's next.

FUN FACT

These galaxies aren't technically "moving away." The space between us and them is expanding. Imagine raisins in bread batter. When you bake the batter, the raisins "move away" from each other because the bread has expanded between them. That's what's happening in space. Therefore, the cosmos is just a giant loaf of bread... and I'm okay with that.

Dark Matter & Dark Energy

I don't have good graphics or images for these as they are... dark, which in astronomy means we can't see them since light doesn't interact with them. So what are they? Simulation errors? Particles from another dimension? Actually visible, but we're not evolved enough to see them? Idk. All we know is that DM and DE exist because of how they affect our Universe.

- **Dark matter** is a hypothetical "invisible" matter that doesn't interact with light but with gravity. Scientists think a matter like this exists because something needs to account for holding galaxies together. If galaxies only contained visible matter, as in what's on the periodic table (things we can see), then stars would be flung off all the time. There must be some extra matter keeping galaxies gravitationally bound. There is much more DM than visible matter, 85% to 15%, respectively. I mentioned CDM (cold dark matter). The "cold" part means that its speed is much slower than the speed of light, which would allow these dark particles to clump.

- **Dark energy** is a mysterious force causing the uniformly accelerated expansion of the Universe (Λ) that scientists know very little about (except for this one fact). The Universe is not only expanding, but something is making it expand faster with time. Gravity is still causing collisions with objects nearby (#Milkomeda), but the farther away you look, the faster those galaxies expand away from us. That doesn't make us the center, however. A being in a distant galaxy would see our galaxy do the same thing! Therefore, nowhere is the center, and everywhere is the center. Trippy!

To sum up: dark energy complies with covid rules, and dark matter doesn't give a damn.

Between all of the possible two trillion galaxies, gas, dust, planets, particles, life, stars... literally anything made of atoms, it's a lot of visible stuff in the Universe. But is it? Not when you compare it to what makes up the rest of our cosmos.

The Universe

- Visible matter ~ 5%
- Dark matter ~ 27%
- Dark energy ~ 68%

Feel insignificant yet?

We pretty much don't know about 95% of the Universe.

The age of the Universe is determined to be 13.7 billion years old, but its diameter is calculated to be 93 billion light years across. How does that make sense? If the furthest we could see is 13.7 billion light years to the beginning of the Universe, how would objects have moved beyond that?

This concept is tough to understand, but it's due to the expansion of space. It's fair to say that space expands faster than light travels. Is everyone lost? Basically, this expansion has caused objects to expand beyond the 13.7 billion light year distance (possibly 2 trillion galaxies).

entire Universe

93 billion light year
observable Universe

As I mentioned, galaxy JADES-GS-z13-0 is at a distance of 33.6 billion light years at the present moment. The light we see left 320 million years after the Big Bang, but over 13.5 billion years since then, it has moved to its redshift distance of 33.6 billion light years!

Imagine someone lets out a massive fart on the street that reaches you a block away. By the time you get to the origin location of the fart, the person has walked away.

Of course, we don't *see* the galaxy at this distance. This is a mathematical calculation based on the Hubble constant. We can only see as far back as the Universe has been around. We are constrained by time, but space doesn't give a damn.

If you remember, the furthest we *can* see is the cosmic microwave background at 380k years after the Big Bang, but its current redshift distance would be about 46.5 billion light years away! It's at the edge of the observable Universe. The white circle in the graphic above represents the CMB. We are alive at a time when we can still see the CMB. In the distant future, those people won't be able to. The CMB is a huge factor in determining the age of the Universe. If our civilization came about billions of years from now, the CMB would have expanded to the point of being undetectable. Could we know the age of the Universe then? Perhaps not!

So what's beyond the observable Universe? Maybe an infinite number of galaxies that we'll never see. All galaxies will be beyond the seeing limit in an unfathomable amount of time. How sad. But good news, we'll all be LONG dead.

End Theories

The four main "end of the Universe" scenarios.

1) **The Big Crunch**: a more cyclical type of Universe as gravity wins and reverses the expansion of the Universe down to what it was like right before the Big Bang. This could lead to an ongoing series of Big Bangs. This isn't the popular idea anymore since we know that the Universe is expanding faster with time.

2) **Heat Death**: currently the most accepted way the Universe will end. Dark energy is causing the accelerated expansion of the Universe. Eventually, everything will separate to a point where galaxies will no longer collide, halting the birthing process of new stars. Basically, everything will die out with a whimper over an unfathomable amount of time. The last to die and evaporate will be black holes.

3) **The Big Rip**: like heat death on crack. Dark energy will play a more significant part and not just keep galaxies apart but will eventually rip atoms apart. It's violent as hell.

4) **Vacuum Decay**: the most fun. The discovery of the Higgs Boson, an elementary particle that gives all matter mass allowed physicists to determine the stability of our Universe. We're in a section called the meta-stability region, meaning we are stable... for now. It's believed that all quantum fields are stable, but the Higgs field, which pervades the entire Universe could be in a false sense of stability (like my career in Hollywood). Some quantum event could ignite its quantum transition and cause it to change its value. The Universe is the way it is because of this field. If it changes, our cosmos gets a new reboot that will travel out at the speed of light! Everything in the path of this reboot will get destroyed, and we'll have an entirely new structure with new laws of physics and chemistry. So we hope the Higgs field is stable because I sure as hell don't want another reboot.

PS, this reboot could have already started somewhere in the Universe and we're just waiting for it to reach us.

Geometry of the Universe

I know I said *geometry*, but don't worry, there's no math. What is the shape of the Universe, and how does it relate to its eventual demise? The shape depends on two things.

- **Actual density**: the amount of mass in the Universe spread over its volume.
- **Critical density**: the amount of matter needed to stop the expansion of the Universe. (Imagine adding too much weight to your squats so you can't stand up.) Enough matter can slow down or stop the expansion.

Ω (Omega) = the ratio of both.

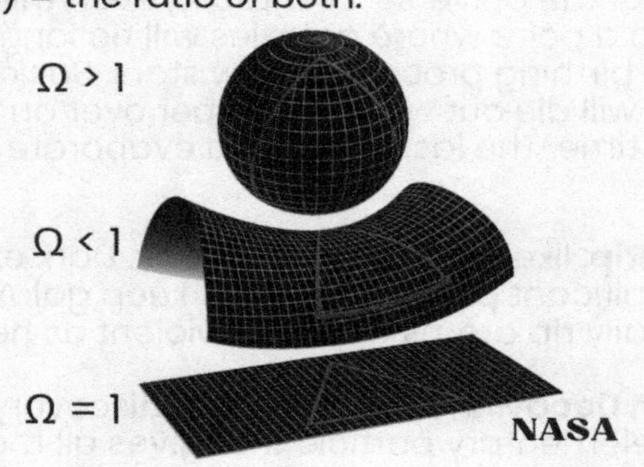

- If the actual density is greater than the critical density, then the Universe will curve like a sphere. This is because there should be enough mass to eventually stop the expansion and contract everything back in (the Big Crunch). This is known as the closed model. The Universe is not infinite but has no "end," like a ball.
- If the actual density is less than the critical density, then the Universe will curve the opposite way like a saddle and expand forever, as there's not enough matter to stop it. This is known as the open model. The Universe is infinite with no bounds.
- If the actual density equals the critical density, as in the Universe is infinite, but there is enough mass to stop the expansion over an infinite amount of time, then we have a flat Universe. This shape leads to heat death.

Based on calculations (by intelligent people), the Universe is considered to be flat!

Thermodynamics

0) 0th Law of Thermodynamics: no joke it's called the 0th law because it was developed after the 1st and 2nd. Thermal equilibrium is the main component and refers to two systems with the same temperature where heat is no longer transferred between them when they come into contact.

1) **Law of Conservation of Energy**: energy cannot be created or destroyed; it can only change forms. Therefore, the total energy content of the Universe remains constant over time. As the Universe expands, the energy of matter and radiation becomes more diluted, but the total energy remains conserved.

2) **Law of Increased Entropy**: energy spreads out, increasing entropy. Entropy is the degree of disorder or uncertainty. Systems tend to go towards higher entropy or disorder as the arrow of time goes forward. Fill a balloon with water and pop it. Crack open an egg. These are simple examples of an increase in entropy. These processes can't spontaneously go backward or decrease entropy. The point is that the Universe works the same way. The death of objects like stars and the expansion of the Universe are leading to an increase in entropy.

3) **Nernst Heat Theorem**: imagine a perfectly ordered crystalline structure in which the constituent atoms, ions, or molecules are arranged in a precise, repetitive, and regular pattern without a single defect. These theoretical structures have a temperature of absolute zero, which results in entropy at zero. Temperature is just the measure of available heat energy or kinetic energies of the motion of molecules. Boiling water is hot because the water molecules are moving faster and faster. Absolute zero (0 Kelvin) is where there would be no motion or kinetic energy. Everything would become solid, and in a perfect system, entropy would be zero as everything is in a maximum state of order. However, absolute zero in our Universe can never be reached in a finite number of steps. Entropy will finally stop increasing when the Universe dies out. It has expanded and cooled to a point where it cannot get any more disordered. Here, we've reached the maximum state of disorder. I look at this as the Universe's goal.

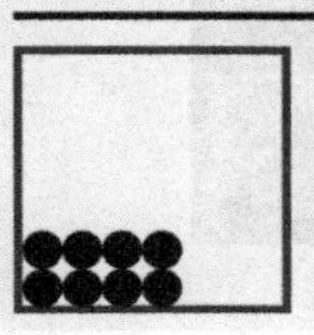
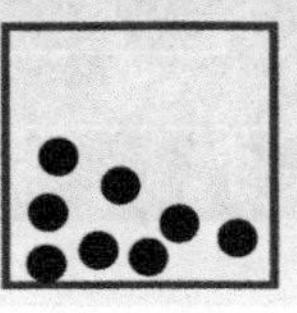
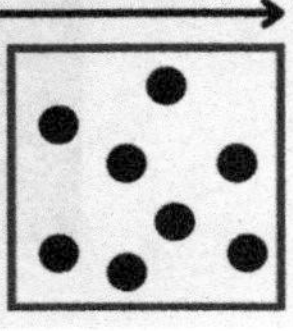

Entropy increases

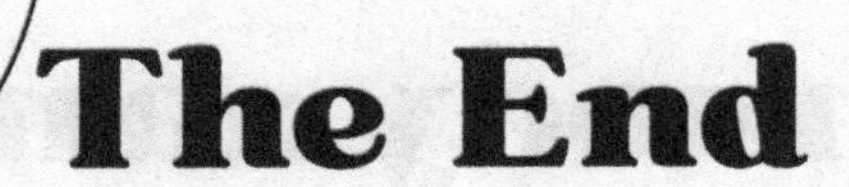

The End

What exactly would be the "end" of our Universe? How is there an end to a space that continues forever?

It's unfathomable for us to think about. Perhaps it's after the last black hole evaporates and the last freely floating photons are heading toward absolute zero as the Universe continues to expand and approach maximum entropy. There is no useful energy available. Matter is gone. Thermal equilibrium has been achieved, as there is no longer an energy transfer, and temperatures have been balanced.

Although this is the Heat Death of the Universe, could it also mirror the moment of the Big Bang?

Only matter experiences and gives value to time, but in the end, there is no more matter. If time and space are the same, and time no longer has meaning, then space in terms of distance no longer exists. We can only cover a distance between two points by using time, but neither points nor time are relevant anymore.

The moment of the Big Bang was also in thermal equilibrium. The only distinction between the Big Bang and Heat Death would be that the latter has experienced time and space... but they are no longer a feature at the end. Therefore, the beginning and end of our Universe are no longer distinguishable.

I wish I had come up with this brain-breaking idea, but I have to credit Nobel Prize-winning physicist Sir Roger Penrose.

If the end of everything gives you angst, see it as no different than the beginning.

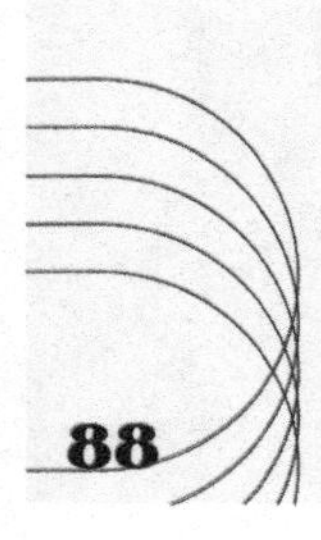

Interstellar

Nebulae

Nebulae are vast clouds of gas and dust. Here are five main types with corresponding pics.

- **Emission Nebulae** are often associated with star-forming regions and are lit up by nearby hot, young stars. They are primarily composed of ionized gas that emits light of various colors. They often appear red or pink due to hydrogen gas.

- **Reflection Nebulae** do not emit light but instead reflect the light of nearby stars. They often appear blue because blue light is scattered most by the tiny dust particles within them.

- **Planetary Nebulae** have nothing to do with planets, weirdly enough. They are the remnants of stars that have shed their outer layers in the late stages of their evolution. These shells of gas and dust often have a spherical shape and symmetrical structures.

- **Dark Nebulae** are dense clouds of dust that block the light from objects behind them. They appear as dark patches against the background of brighter stars or nebulae.

- **Supernova Remnants** are the remnants of massive stars that have exploded as supernovae. The explosion scatters the star's outer layers into space, forming a shell of gas and dust.

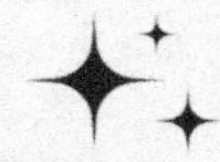

WTF is a Black Hole?

Because of all this packing, their gravity is so strong that not even light, the fastest thing in the Universe, can escape from it.

I always give the trampoline analogy. Imagine the fabric of spacetime is the fabric of a trampoline. Put a golf ball, a basketball, and a bowling ball on it. Or put your fat ass on it and compare it to the other balls. Which object has the largest "dip" in the trampoline? Probably your fat ass. That dip is the gravity of the object.

I know I said earlier that you shouldn't think of a black hole as an actual hole, but since the trampoline is 2D, we now have to. Imagine a hole in that fabric. Whatever you send towards it will fall in and won't come out. Wouldn't you say the hole's "dip" is the greatest among the objects? (More on the "dips of gravity" later.)

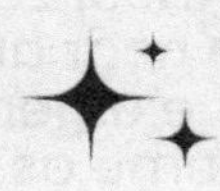

Since not even light can escape from one, we don't actually see them. We see their effects on surrounding matter. However, what if there isn't any surrounding matter? Then they lurk... undetected.

Black Hole Structure

Singularity: at the center where the immense mass resides. It's a point with a minuscule volume and essentially an infinite density. The laws of physics are thought to break down here!

Event Horizon: a pretty decent horror movie from the 90s, AND the point of no return around the black hole. It's a sphere where, once you enter it, only something faster than light can escape. So basically, nothing we know of can escape it. Once this area is crossed, the matter gets squished into the singularity.

Shwarzschild radius: not really a part of the "structure," but is an integral part of the definition of a black hole. It's the distance from the concentration of mass to the event horizon. This radius is where the mass has an escape velocity that equals the speed of light. If you launched a rocket from this object, it would need to travel at the speed of light to escape the mass's gravity. So, if you're an object smaller than your Schwarzschild radius, you're a black hole because your escape velocity is now > than the speed of light. For example, knowing the Sun's mass, it would have to be squished within a radius of about 1.8 miles (2.9 km) to be a black hole. At this size, its gravity would be so strong that light couldn't escape. If light CAN escape you, you're not a black hole. This radius lets us know what it would take for light not to be able to escape from you based on your mass.

event horizon
Shwarzschild radius
singularity

Accretion disk: not a part of a black hole, but this term comes up sometimes. They are flattened disks of rapidly rotating gas and particles steadily falling toward the gravitational force. Accretion is the increase in mass of a celestial object by gravitational attraction. As in, even space objects gotta eat!

Hawking radiation: proposed by Steven Hawking, it theorizes that subatomic pairs of particles that naturally arise near an event horizon split, with one falling into the BH and the other escaping. This could imply that BHs evaporate over time as they emit this weak radiation.

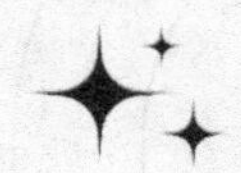

Black Hole Types

Scientists categorize black holes into three types:

1) **Stellar-mass**: these are the ones that are caused by a star with mass > 20 times the Sun's mass before it went supernova. These newly born black holes can range from a few to hundreds of times the Sun's mass. So yes, the mass before and after don't have to match, but they are correlated.

- Can gain mass through collisions with other objects.
- So far, 50 have been found in a binary system and are paired with another star.
- About 100 million are estimated to exist in our galaxy.
- In 2022, a rogue black hole was discovered. It wanders our galaxy alone. (Imagine how difficult the process of finding that one was!)
- At the time of this writing, scientists announced that a few of these are predicted to exist only 150 light years away in the Hyades star cluster, making them the closest known black holes.

2) **Supermassive**: at the center of most galaxies and range in mass from hundreds of thousands to billions of times the mass of the Sun!

- Their origin isn't fully known.
- Can grow by feeding on surrounding material (accretion).
- Can merge with other supermassives to grow bigger.
- Largest one so far is TON 618 with a mass 66 billion times the Sun! (Look this one up, it's crazy!)

3) **Intermediate-mass**: predicted to range from a hundred to hundreds of thousands of times the Sun's mass. BUT...

- No candidates have been officially confirmed.

4) **Primordial**: throwing in this category for fun as these are only theorized to exist.

- Formed in the 1st second after the Big Bang.
- Mass can range from 100,000 times smaller than a damn paperclip to 100,000 times more than the Sun.

Black Hole Breakthroughs

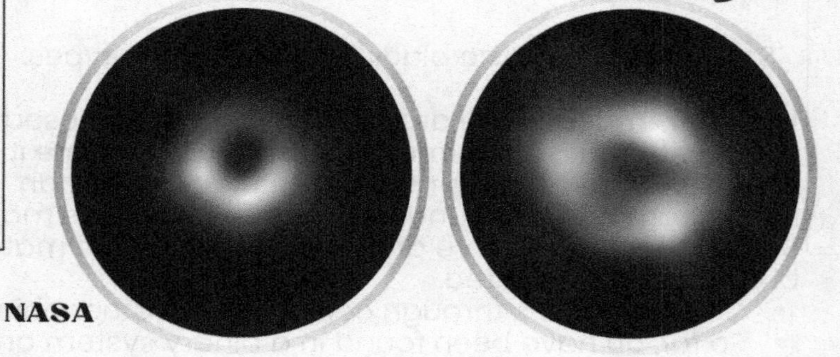

NASA

The orange donut on the left is the first real image of a black hole. If that doesn't blow your small mind I don't know what will. It's of the supermassive black hole at the center of the M87 galaxy located 55 million light years away and is 6.5 billion times the mass of the Sun. Just recently (September 2023), scientists confirmed that it spins.

The image was taken by an international network of radio telescopes collectively known as the Event Horizon Telescope. By using radio teles from around the world they essentially created an aperture the diameter of the Earth! See how collaborative science can be?

We say the first image of a black hole, but I'm hoping you're asking, "How do we take a pic of something that consumes all light?" If you did, you're smart. If not... there's still some learning to do. Technically, it's an image of the black hole's shadow or silhouette against the accretion disk. The image on the right is of our supermassive black hole, Sagittarius A * released in 2022! It's 4 million times the Sun's mass.

M87 vs Sag A *

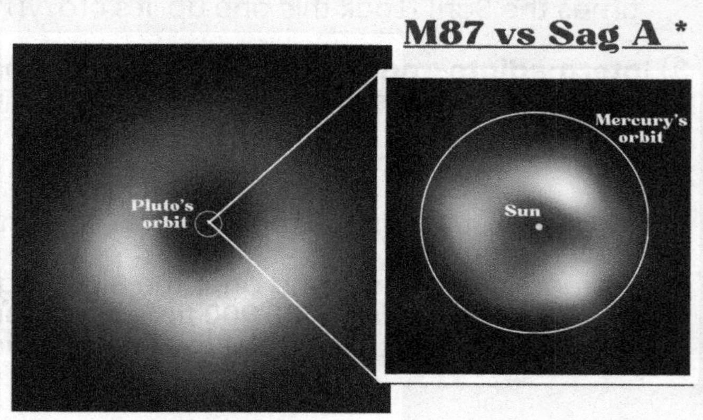

Quasars

Quasars, aka quasi-stellar radio sources, are extremely energetic and luminous objects found at the centers of various galaxies and are thought to be powered by supermassive black holes. They are among the most powerful and active objects in the known Universe!

Most quasars emit powerful jets, which are a stream of charged particles. When a quasar's jet points directly at Earth, it's known as a blazar (not blazer).

Scientists theorize that energy emitted by quasars is due to the accretion of mass onto the supermassive black holes at the centers of galaxies. As these black holes pull surrounding material in, the swirling disk heats up and emits large amounts of radiation from radio waves to X-rays. So, while black holes emit no light, quasars emit a lot! To sum up, a quasar is a high-energy phenomenon that is thought to be powered by a supermassive BH at the center of a galaxy.

As their high redshifts indicate, most quasars are billions of light years away.

Some believe they are "activated" when two galaxies merge, allowing the supermassive black holes to merge and feed on the new chaos.

AGNs:

- The galaxy's central region, where a quasar is found, is known as an active galactic nucleus (AGN).
- They are incredibly bright as they can emit radiation from across the EM spectrum and often outshine entire galaxies. Their luminosities can be thousands to millions of times greater than the Milky Way!
- Quasars are a specific type of AGN.
- Seyfert galaxies are another type. Both represent the two largest groups of active galaxies. However, Seyfert galaxies aren't as bright, are much closer, and are often found in spiral galaxies.

Quasars are valuable probes for studying the early Universe, galaxy evolution, and supermassive black holes.

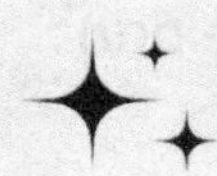

Gravitational Waves

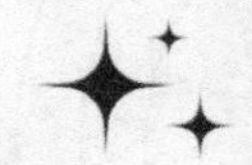

These things are fucking trippy. Einstein first theorized their existence in 1916. Then about 100 years later, we fucking detected them! Gravitational waves are the ripples of the fabric of spacetime itself caused by massive objects accelerating and interacting in some way, orbiting each other, colliding, blowing up, etc. The more cataclysmic the event, the stronger the waves. These waves then ripple out at the speed of light. The first detection was of a violent collision between two supermassive black holes 1.3 billion years ago (1.3 billion light years away). The waves generated by that event were picked up by the sensitive instruments known as LIGO in 2015.

Can I say how amazing it is that humans are capable of building instruments that can measure a change in distance 1/10,000th the width of a fucking proton?! You should go and watch some videos on how LIGO works. I'll wait.

The better we get at detecting these, the further back in time we can study, even going back well before the CMB. Unlike light, GWs aren't constrained by the early Universe's processes. One day, we might be able to map what the Universe was like; get this: 10^-32 seconds after the Big Bang! Is that tiny of a number even fathomable?

It'll be a while before we can do that, but we have made strides in substituting pulsars for LIGO. These guys are very timely space lighthouses. If we can detect that their energy jets get thrown off by even a fraction of a second, we can measure the GWs that cause those disruptions. An instrument the size of the Milky Way, anyone?

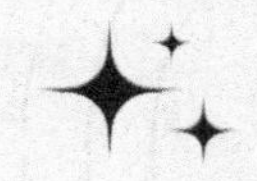

Gamma-Ray Bursts

Gamma-ray bursts (GRBs) are short-lived bursts of gamma-ray light, which is the most energetic form of light as you know from the EM spectrum. They can outshine an entire supernova and are hundreds of trillions of times brighter than the Sun. Gamma rays are dangerous but our ozone and atmosphere help block them out.

- GRBs release more energy in 10 seconds than the Sun will release over its 10 billion-year lifespan.
- Their sources were a mystery until recently.
- There are two kinds: short duration and long duration.
    - Short durations last from a few milliseconds to 2 seconds.
        - It is caused by the merger of two neutron stars or a neutron star with a black hole.
    - Long durations last from 2 seconds to several minutes.
        - Caused by a massive star going supernova, though not all supernovae cause them.
    - The processes that create short and long-duration bursts result in a new black hole.
- A GRB in our direction could evaporate our ozone layer.
- If a GRB occurred within 200 light years, it would fuck us up by vaporizing us, but no stars in that distance are candidates.
- GRBs are detected almost daily and come from all over the Universe.

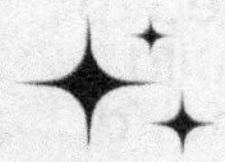

Fast Radio Bursts

Fast radio bursts (FRBs) are intense pulses of radio waves that emit more energy in a fraction of a second than the Sun emits over 100 years.

- Discovered in 2007.
    - Only about a few dozen have been discovered.
    - Farthest one came from 8 billion light years away.
    - Their origins are still unknown.
    - They come from all over the Universe.

- Repeating FRBs were discovered in 2016.
    - The pulses were random and sporadic.
    - About 28 were discovered using the radio telescope known as CHIME in Okanagan Falls, Canada, between September 2018 and October 2019.
        - One known as 180916.J0158+65 repeats regularly at 16.35 days. Coming from a galaxy 500 million light years away, the source sends out 1-2 bursts every hour for 4 days, then goes silent for 12 days. The process then repeats.
    - Candidates for repeating FRBs could be neutron stars, but the sporadic and elongated pulses of the waves don't tend to match how rapidly neutron stars spin.
    - The regular pulses of the one with the crazy name can help scientists find a source. Perhaps it is a neutron star that orbits another star. Maybe the signal gets obscured when it's behind its companion star.
    - Patterns are helpful because we see patterns in space - rotations, orbits, etc.

- Earlier, I mentioned that the missing baryonic (visible) matter from galaxies may have been dispelled into the space between them.
    - FRBs can interact with this diffuse (spread out very thinly) matter with their long wavelengths and give us an inventory of how much matter there is.

- Some speculate that it's alien in origin, but those chances are very slim because the objects needed to emit them would be planet-sized. If they could build that, we could find them by all the other things they'd be capable of building, and we haven't.

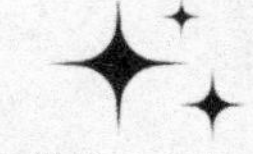

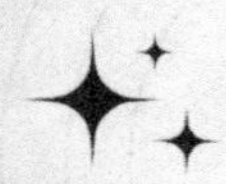

LFBOTs

Luminous Fast Blue Optical Transients, or LFBOTs, are a rare breed of short astronomical events that are still a mystery to scientists. Only a handful have been discovered. They shine intensely in blue light for only a matter of days before they diminish.

- Discovered in 2018.
  - The first one is known as AT2018cow.
  - Took place in an arm of a spiral galaxy 200 million light years away.
  - Up to 100 times brighter than a regular supernova.
  - LFBOTs have been seen about once per year since 2018.
  - Nicknamed after animals based on their last three letters.
  - Others are Camel, Koala, and Tasmanian Devil.

- Latest one was detected in April 2023.
  - Named AT2023fhn or Finch.
  - 3 billion light years away.
  - The brightest one yet.
  - The first one detected that wasn't inside a galaxy!
  - This explosion occurred in the space between two galaxies, adding to the mystery.
  - Can rule out a traditional supernova since stars big enough to do so wouldn't live long enough to make it out that far from their original host galaxy.
  - The leading explanation is the merger of two neutron stars that got flung out, resulting in a kilonova, which is 1,000 times brighter than a supernova.
  - Temperature measurements put it at 36,000°F (20k°C).

- How they differ from supernovae.
  - Last for only a few days instead of weeks or months.
  - Much brighter.
  - Not as hot as supernovae, which can reach millions of degrees.
  - Supernovae are explainable by massive star deaths, but Finch took place in a void.

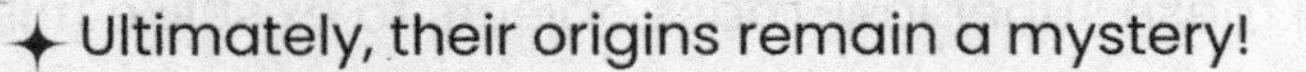

Ultimately, their origins remain a mystery!

JuMBOs

Once again, at the time of writing this book, a discovery has been made! Jupiter-mass binary objects (JuMBOs) have perplexed scientists because they don't know what these guys are!

JWST looked at the Orion Nebula 1,334 light years away, a stellar nursery below Orion's Belt that's visible to the naked eye. However, it found some interesting objects that might herald a new astronomical category.

It found 40 pairs of Jupiter-mass objects that weren't revolving around any stars.

They're too small to be stars but don't fit into what it means to be a planet since they aren't orbiting any stars. (Maybe they're rogues?)

The weirdest part... why are they in pairs? As you can see in the image to the left, each box shows two dots representing the JuMBO pairs.

Gas physics (yes, that's a thing) says these guys shouldn't be able to form here.

Basically, the whole thing is weird!

What we do know about them:

- Large, gassy, and hot objects.
- Atmospheric analysis reveals methane and steam.
- About 1 million years old.
- Temps range from 1k°F - 2,300°F (537 - 1,260°C).
- Will eventually cool off, but aren't contenders for habitability since there's no surface liquid water.
- Their masses range from 0.6 - 13 Jupiter masses.
- Pairs take between 20k - 80k years to orbit each other.

Planets are always getting ejected from their solar systems. Still, the fact that many of these are in pairs that stayed together is throwing scientists for a loop, as existing theories of star and planet formation never predicted these super low-mass objects to exist like this!

Also, definitely google the JWST images of the nebula!

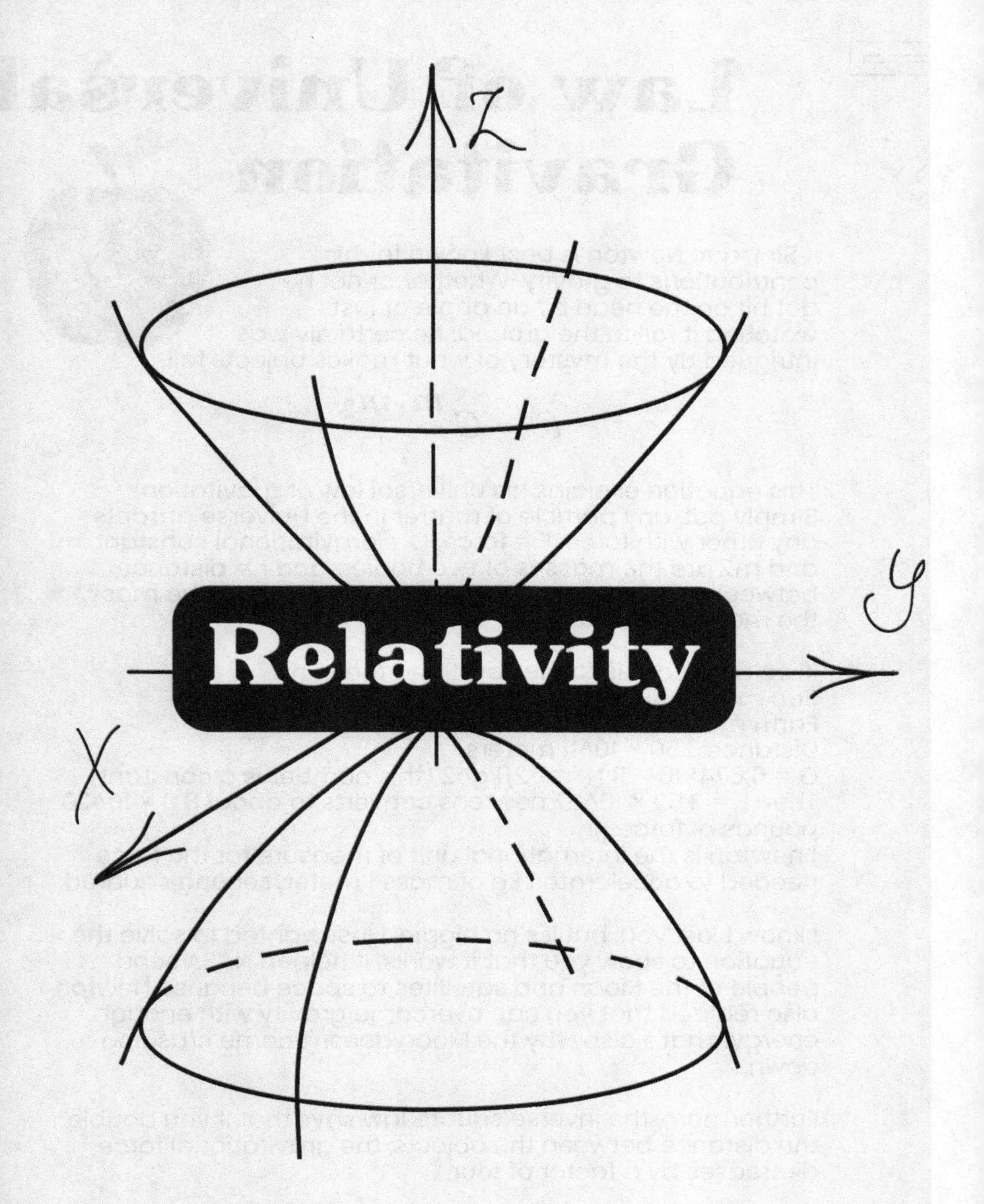
Z
y
Relativity
X

Law of Universal Gravitation

Sir Isaac Newton is best known for his contributions to gravity. Whether or not he got hit on the head by an apple or just watched it fall to the ground; he certainly was intrigued by the mystery of what makes objects fall.

$$F = G\frac{m_1 m_2}{r^2}$$

This equation explains his universal law of gravitation. Simply put, any particle of matter in the Universe attracts any other with force. F = force, G = gravitational constant, m1 and m2 are the masses of two bodies, and r = distance between the center of those masses. The larger the mass, the more its gravitational attraction.

If we applied this to the Earth and the Sun:
Sun mass: 1.99 × 10^30 kilograms
Earth mass: 5.98 × 10^24 kilograms
Distance: 1.50 × 10^11 meters
G = 6.674×10^-11 N · m^2/kg^2 (this number is a constant)
Then F = 3.52 × 10^22 newtons converts to about 8.0 × 10^20 pounds of force.
1 newton is the international unit of measure for the force needed to accelerate 1 kg of mass 1 meter/second squared.

I know I lost you, but it's no biggie. I just wanted to solve the equation to show you that it works. It helped NASA send people to the Moon and satellites to space because Newton also realized that you can overcome gravity with enough energy. That's also why the Moon doesn't come crashing down.

Furthermore, the inverse square law says that if you double the distance between the objects, the gravitational force decreases by a factor of four.

Your weight factors in gravity. Your mass will always be the same, but how much you weigh is your mass x gravity. That's why you're even more of a fat ass on Jupiter since its gravity is stronger than Earth's.

Relativity

However, Newton's gravity law and classical mechanics paint only part of the picture.

Classical mechanics is the branch of physics that deals with the motion of objects under the influence of forces.

It is highly successful in describing the motion of everyday objects, such as projectiles, planets, and pendulums. It tends to fail in very extreme or very precise conditions like speeds close to the speed of light, gravity near a black hole, gravity in the early moments of the Universe, explaining the precession, or slight change in Mercury's orbit around the Sun over time, and in quantum mechanics. That's where Albert Einstein's theory of relativity (minus quantum mechanics) comes in.

Science has obviously improved since Newton's time in the 1600s. When Einstein's time came along centuries later, experiments and observations that could show where classical mechanics didn't work would be conducted.

In the early 1900s, Einstein developed his theory of relativity, which consists of Special and General Relativity. Special deals with light speed and general deals with gravity. It's easy to remember because *Special* starts with *S* and talks about speed, while *General* starts with *G* and talks about gravity.

It's called "relativity" for a reason. Einstein gave us the idea that the laws of physics are relative and depend on the motion of observers. It's almost like he gave more power to the individual than the collective. I'll elaborate on SR and GR more in the following few pages.

Overall, Einstein's development of the theory of relativity was driven by a desire to reconcile the fundamental principles of physics, provide a more accurate description of the physical world, and address the experimental and theoretical challenges posed by classical physics. His work in this area had a profound and lasting impact on the field of physics, leading to a new understanding of space, time, and gravity. As you'll come to learn about the "theory of everything" and its importance, know that even Einstein tried to find one... but ultimately couldn't.

Special Rel

Before this, it was thought that time was constant and the speed of light could change. Some of the key motivations for SR were observations and experiments like Michelson-Morley, which suggested that the speed of light in a vacuum is constant and the same for all observers, regardless of their motion. This contradicted classical mechanics, which assumed that velocities were added linearly. That's when Einstein popped out SR.

Special Relativity came in 1905. This is where the famous E=mc^2 equation comes from. SR states that as an object approaches the speed of light, its mass becomes infinite and so does the energy required to move it. Thus, the speed of light is the cosmic speed limit. This would now have to make time malleable.

If I were on a train moving 50mph (80 km/hr) and threw a ball at 30mph (48 km/hr), I would see the ball move at 30 mph, but a person outside the train watching me (creepy) would see the ball go 80 mph (50+30). It's all relative, get it? However, the speed of light is the same, regardless of your frame of reference. If I were on that same train and shot a laser, I'd see the laser's light move at the speed of light... and so would the creep outside. But based on the last example, why wouldn't we add the train's speed to the speed of light? We can't. This would break the cosmic speed limit; therefore, time would slow down for me on the train without me knowing *relative* to the person outside. The cosmic speed limit is a speed, which is distance divided by time. If the speed goes up to *relativistic* levels (speeds at a significant fraction of the speed of light), your time will slow down. It's fucking crazy... and proven!

Astronaut Scott Kelly is a twin. When he was on the ISS traveling at 17,500 mph (28,163 km/hr), time slowed down for him relative to his brother on Earth. It's a small change since 17k is slow as hell compared to the speed of light, but he would become 25 microseconds a day younger than his brother. This is known as time dilation—more on the next page.

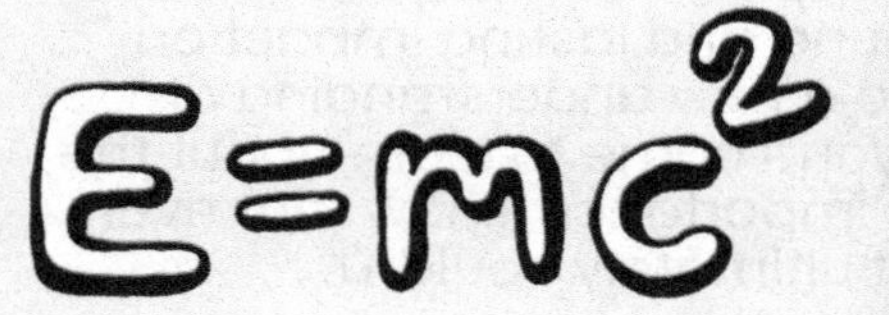

Energy = mass x the speed of light squared.

Energy and matter are different forms of the same thing.

Multiply the mass and the speed of light to find out how much energy is contained in the mass.

Under the right conditions, energy can become mass and vice versa.

Time Dilation

Time moving slower for you the faster you go is known as time dilation. The equation may look scary, but it's actually pretty easy to solve. So just for fun let's do it.

t' - time measured by observer
t - time measured by traveler
v - speed of traveler
c - speed of light

$$\Delta t' = \frac{\Delta t}{\sqrt{1 - \frac{v^2}{c^2}}}$$

The parameters: a 15-year-old is sent into space for 5 years, traveling at 99.5% the speed of light. When she returns to Earth, she will be 20 years old, BUT how much time has passed for everyone on Earth?
NOTE: 5 years = 157,680,000 seconds, and C = 1

$$t' = \frac{157,680,000 \text{ seconds}}{\sqrt{1 - \left(\frac{.995c}{c}\right)^2}}$$

$$\sqrt{1 - (.995)^2}$$

$$\sqrt{1 - .990025}$$

$$\sqrt{.009975}$$

$$\frac{157,680,000}{.09987}$$

$$= \frac{1,578,852,508}{31,536,600}$$

(number of seconds in a year)

answer = 50 years!

If you can read my handwriting, everyone on Earth would have aged 50 years while the girl only aged five. Crazy stuff!

General Rel

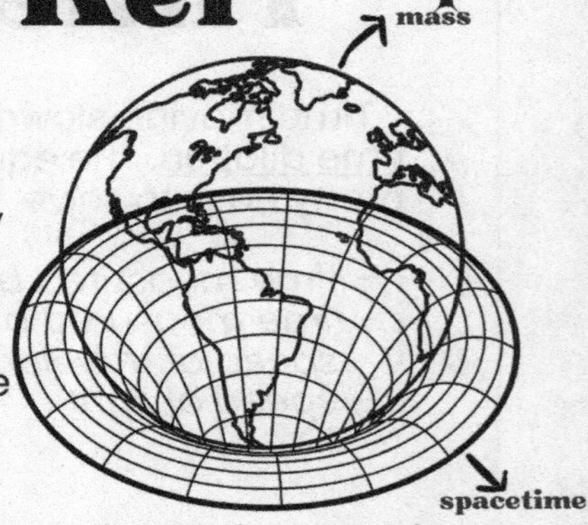

General Relativity came in 1915 and is an extension of Special Relativity. It explains gravity as a curvature of spacetime caused by mass and energy rather than a force between two objects. Massive objects, like planets or stars, warp the fabric of spacetime around them, causing other objects to move on curved paths.

GR also has time dilation. Objects experience time running slower when they're in stronger gravitational fields. Some sci-fi films and shows have demonstrated this well, and in the real world, it has been measured. The core of our Earth is 2.5 years younger than the surface! The deeper you are in an object's gravity well, the slower time moves for you. Remember, space and time are the same thing. If an object's mass distorts the fabric of space, it will also distort time. Even GPS satellites have to account for this because clocks on Earth will run slightly slower than clocks on the satellites! It's only a difference of microseconds, but that can add up, and I'm not trying to sit in LA traffic more than I have to.

GR predicted the existence of gravitational waves. If only Einstein had been around for this discovery. But considering how much he valued imagination, he would have "seen" them in his mind anyway. FYI, my name, Kalpana, means *imagination* in Sanskrit. Just sayin'.

GR was a revolutionary way to see our Universe. It shows you that even though objects in space are far away, they're still connected by this weird fabric.

Einstein's field equation for GR

$$G_{\mu\nu} + \Lambda g_{\mu\nu} = \frac{8\pi G}{c^4} T_{\mu\nu}$$

Matter tells spacetime how to curve.

Spacetime tells matter how to move.

SR and GR give us the secret to living a long time. Get as close to the Earth's core as possible and move really fast. Anyone wanna be a mermaid?

Gravitational Lensing

GL is a very trippy phenomenon that happens when objects like galaxies appear distorted due to their light bending around foreground objects.

NASA/JWST

As you can see in this image, which was the first image released by JWST in July 2022 of the galaxy cluster SMACS 0723, some of the galaxies seem to curve. They appear weirdly misshapen and distorted. Some even repeat!

This is due to gravitational lensing, further proving Einstein's GR theory. Since gravity results from the curvature of spacetime by massive objects, light will bend along that curved path.

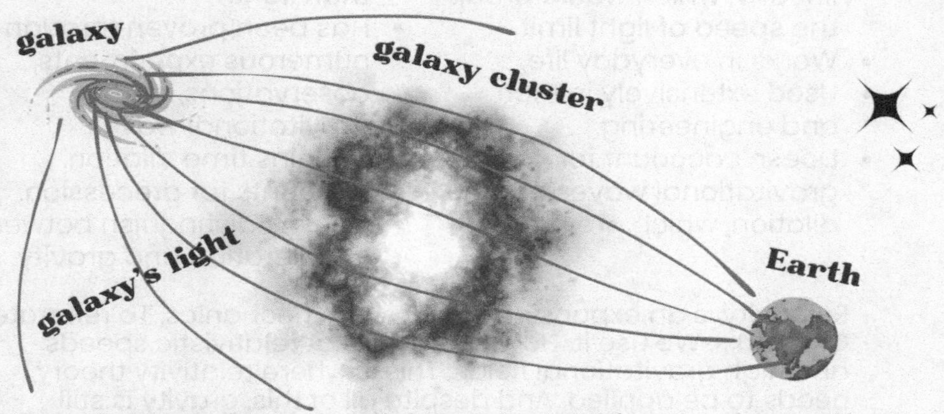

As you can see by my crudely made graphic, a background galaxy's light will bend around a large foreground object's curvature of the fabric of spacetime. This is so cool because it allows us to see objects that would otherwise be obscured!

Newton vs. Einstein

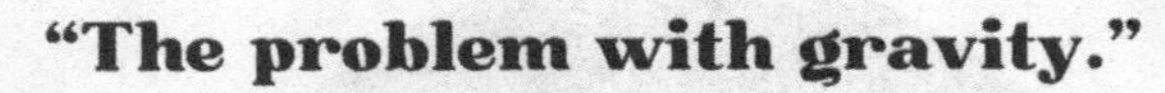

"The problem with gravity."

Newton

- Gravity idea introduced in 1687.
- Part of classical mechanics.
- Gravity is a force expressed mutually between two objects in relation to their masses.
- Gravity is a "pull."
- Limited to speeds much slower than the speed of light, and lower gravity fields like solar systems.
- Gravity is an instantaneous force.
- Assumes an absolute and independent space and time.
- Velocities are added linearly, which would break the speed of light limit.
- Works in everyday life.
- Used extensively in tech and engineering.
- Doesn't account for gravitational waves or time dilation, which are proven.

Einstein

- Gravity Idea introduced in 1915.
- Part of relativity theory.
- Gravity is a curvature of spacetime proportional to an object's mass.
- Gravity is a "push."
- Expands on Newton's equations to include speeds near or at the speed of light (relativistic speeds) and higher gravity fields.
- Gravity is not instantaneous as this violates relativity.
- Shows that space and time are one and it's all relative to the observer.
- Nothing can travel faster than light.
- Has been proven through numerous experiments, observations, and gravitational waves.
- Explains time dilation.
- Accounts for precession.
- Doesn't distinguish between acceleration and gravity.

Relativity is an expansion of classical mechanics. To reiterate, CM works. We use it. However, it fails at relativistic speeds and high gravitational fields. This is where relativity theory needs to be applied. And despite all of this, gravity is still confusing as hell, and we don't even have a complete picture yet! Neither works in extreme situations like the singularities at the center of black holes. We still need a piece of the cosmic puzzle, which is the next topic.

Quantum

The Atom

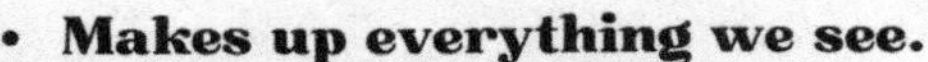

- Makes up everything we see.
- Has a radius roughly 1.181×10^{-10} inches.
- Opposite charges attract while the same charges repel.
- Is mostly empty space but includes...

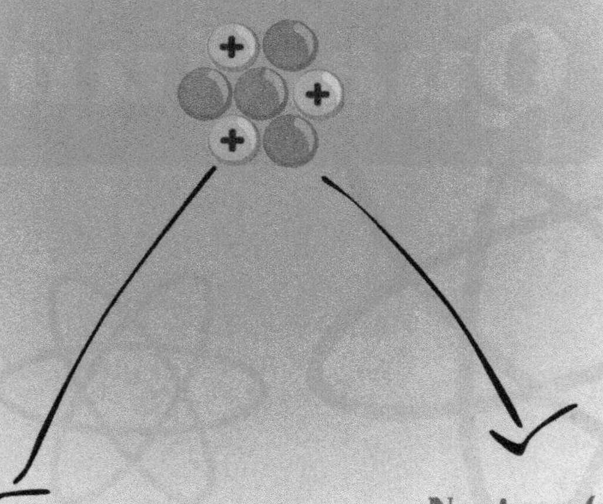

- Electron (e^-).
- Negative charge.
- Cannot be broken down further (elementary particle).
- Smallest mass of the three.
- Exist in the electron cloud.
- Orbit in "shells" around the nucleus.
- Locations in the shells are based on probability.

- Proton (p^+).
- Positive charge.
- Made up of quarks (type of elementary particle).
- Similar mass to neutron but slightly smaller.
- Make up the nucleus of the atom.

- Neutron (n0).
- Neutral charge.
- Made up of quarks (type of elementary particle).
- Largest of the three.
- Make up the nucleus of the atom.

This may not look like the traditional atom we all saw in school, but the quantum mechanics model is the most accurate model of the atom so far.

Quantum (fml)

This subject is one of the most fascinating yet one of the most headache-inducing because things at this scale behave so differently than what we observe in our everyday world.

Quantum mechanics deals with the microscopic: atoms, subatomic particles, light, etc. while classical mechanics and relativity deal with the macroscopic: planets, solar systems, even motions on Earth, etc.

As our understanding of the tiny world improved, scientists in the early 1900s realized that if you applied CM to an atom, the physics wouldn't work as the electron(s) would spin inwards causing the atom to collapse. The solution was to think of electrons as waves instead of particles. This is where we get a new equation called the wave function (instead of using the famous f=ma equation). Wave functions "collapse" to become particles.

There is complete certainty in CM, while there's uncertainty in QM. The Heisenberg Uncertainty Principle states there is an inherent limit to how precisely certain pairs of properties, like position and momentum, can be simultaneously known for particles. The more precisely the position is known, the more uncertain momentum is, and vice versa. This introduces fundamental uncertainty into measurements.

To sum up, quantum mechanics is about probability, while classical mechanics is about certainty. It's weird for us not to know exactly how a system will turn out if we know its components. I know I said that's not how we observe things in our everyday lives... but it also is. We don't know how our lives will turn out, but we can sometimes predict aspects based on probability. For instance, moving to Hollywood, working on my craft, and networking yields a higher probability that I'll "make it" than if I had just stayed in Ohio and expected the same results. The first case isn't a guarantee either, but it yields a higher probability than if I didn't move. Classical mechanics would say that if those three things are all you need to "make it," then you will with absolute certainty. So, even though QM is observed to be strange, I'd say it's a closer metaphor for life.

The following pages describe the Copenhagen Interpretation of QM.

QM History

I find QM history to be fascinating, so here are SOME of the key moments. Remember, QM came about because CM wasn't working at the tiny scales, so these nerds had to find other explanations.

Planck's Quantum Hypothesis (1900):
- Max Planck introduced the concept of quantization. He proposed that energy is quantized in discrete units or "quanta" instead of just any amount. Remember electrons needing a specific amount of energy to jump levels? Picture a faucet dripping water instead of streaming water. This was the birth of quantum theory.

Einstein's Explanation of the Photoelectric Effect (1905):
- In the photoelectric effect, light striking a metal surface ejects electrons as the photons give the electrons enough energy to break free from the atoms. This showed us the concept of photons as discrete packets of energy, thus proving Planck's point.

Bohr's Model of the Hydrogen Atom (1913):
- Niels Bohr combined classical mechanics with quantization principles to explain how electrons behaved in Hydrogen atoms. They orbit the nucleus like planets orbit the Sun and can only move "up or down" a shell depending on how much energy they gain or lose. This was a transitional step as it didn't accurately describe how electrons "orbit."

De Broglie's Wave-Particle Duality (1924):
- Louis de Broglie proposed that particles, like electrons, exhibit particle-like *and* wave-like behaviors. He introduced the idea that particles have associated wavelengths, thus unifying the concepts of particles and waves and laying the foundation for wave mechanics. (Particle-wave duality explanation to come).

Schrödinger's Wave Mechanics (1926):
- Erwin Schrödinger formulated wave mechanics and introduced the Schrödinger equation. This equation describes the likelihood of a particle's location (or state) in a system. Electron orbitals, where electrons are likely to be found, are based on probability distributions.

Quantum Mechanics Consolidation (1927-1930s):
- QM became a complete and consistent framework for describing the behavior of matter and energy.

Development of Quantum Field Theory (1930s):
- QFT combines special relativity and quantum mechanics. It describes the Universe as having invisible fields that permeate its entirety. These fields can vibrate and interact with each other. The energy in these fields gives rise to their corresponding particles: photons, electrons, Higgs bosons, etc. It is one of the most successful theories in physics.

Atom Bomb

Robert Oppenheimer created something that changed humanity forever.

In 1938, German physicists Otto Hahn and Fritz Strassmann and Austrian physicist Lise Meitner discovered nuclear fission, where the nucleus of an atom gets split into two or more smaller nuclei. They realized that this process released a large amount of energy.

FUN FACT

Lise Meitner had to escape Nazis in Germany and was the one who figured out that they had successfully split the atom from afar; however, she was never given credit nor the Nobel Prize, while the other two men did. This was, unfortunately, a common occurrence for women in STEM.

In the 1940s, Enrico Fermi used a controlled setup to demonstrate that nuclear fission could be sustained and controlled, leading to a self-sustaining nuclear reaction.

The development of methods for separating isotopes of uranium and plutonium was essential. These isotopes were the primary fuel for atomic bombs.

In 1943, The Los Alamos Laboratory in New Mexico, a top-secret site for the development of nuclear weapons under the leadership of Oppenheimer, brought together a large team of scientists and engineers to work on the bomb's development. This was known as the Manhattan Project.

In July of 1945, the Trinity Test was the first successful detonation of an atomic bomb and confirmed the feasibility of the implosion-type design.

In August 1945, the atomic bombs "Little Boy" (uranium-235) and "Fat Man" (plutonium-239) were dropped on Hiroshima and Nagasaki, Japan, leading to the end of World War II.

Oppenheimer did realize the implications of what they had created. He advocated for international control of nuclear weapons and the peaceful use of atomic energy. Although it blew humanity (pun intended) into a new era (good or bad), it showcased what a collaboration of some of the most brilliant minds can do.

Four Forces

These four fundamental forces explain every interaction in the Universe listed from weakest to strongest.

- **Gravity**: hopefully, you better understand this and its complexities now so I won't delve into it again. Of the four forces, it's the only one that applies to the large scale and is the weakest. We overcome it here on Earth.

- **Weak force**: aka the weak nuclear force is responsible for particle decay. This is when a subatomic particle changes into another. These particles can only decay when they are incredibly close to each other. Furthermore, they can only decay into lighter particles as per the law of the conservation of energy. It's not favorable for energy to be required to "decay up." For example, a free neutron can decay into a proton since the latter is lighter. This force is responsible for the nuclear fusion that takes place inside stars and for carbon dating.

- **Electromagnetic force**: the most understood force consisting of the electric and magnetic forces. This one binds negatively charged electrons to positively charged atomic nuclei. It's responsible for friction, drag, and elasticity and holds solid shapes together. Separately, it holds balloons to a wall and magnets to a fridge. Together, they give rise to electromagnetic waves comprising the EM spectrum. Did you know that you technically never touch anything? The electrons in your skin repel the electrons of an object you're holding (- repels -, + repels +). What you feel is the force. So, may the force be with you.

- **Strong force**: aka the strong nuclear force is the strongest of all 4 forces. Some information says it's 6 thousand trillion trillion trillion times stronger than gravity! This goes on an even smaller scale than the weak and electromagnetic forces because it binds together the subatomic particles like quarks, which make up protons and neutrons. Some of the force also binds the protons and neutrons together.

Once upon a time, these forces were unified as one. They began to split within fractions of a second after the Big Bang. Gravity first, then strong, and then "electroweak" split last. When they were unified, think of just 1 fundamental force that did everything, like a dentist (hi, mom!) who also cleans your teeth and does your braces. When they "split," they assumed distinct characteristics and duties, like having a hygienist and orthodontist to clean your teeth and do your braces instead.

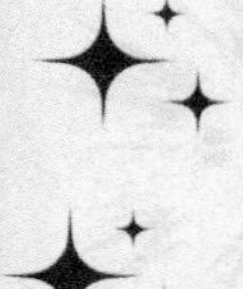

Standard Model

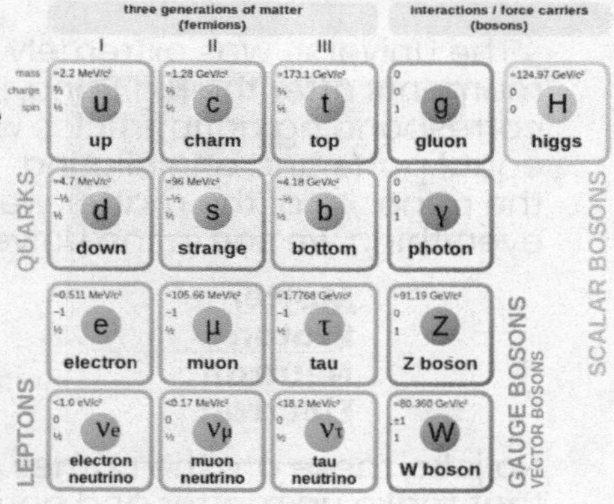

wikimedia commons: MIssJ

These words look made up, don't they? This standard model is the theoretical framework of particle physics that describes the fundamental building blocks of the Universe and their interactions. The four forces come in handy here.

- **Quarks**: the building blocks of protons and neutrons. They come in 6 types. These experience all four forces. Since they experience the strong force that binds atomic nuclei, they can't live freely outside an atom. The force particle that binds quarks is the gluon, a type of boson. Bosons are force-carrying particles.

- **Leptons**: these guys don't experience the strong force so they can live freely outside of an atom (although not for long). Electrons make up the outer shells of atoms and have a negative charge. Muons and taus are created through high-energy particle interactions and decay processes. Neutrinos are much lighter than electrons and have a neutral charge. They weakly interact with matter and are hard to detect. Each lepton has a corresponding neutrino, the only particle that can oscillate between being an electron, muon, or tau neutrino. The force particles that govern their weak interactions are the w and z bosons.

- **Bosons**: photons describe the *electromagnetic force*. The Higgs boson is the particle that gives all matter mass. The discovery of it in 2012 was a huge fucking deal and acted as a confirmation of the standard model.

Notice how the only fundamental force that's missing is gravity! That's why the search for a quantum explanation for gravity is essential.
Side note: it's also missing particles to describe dark matter and dark energy.

Antimatter

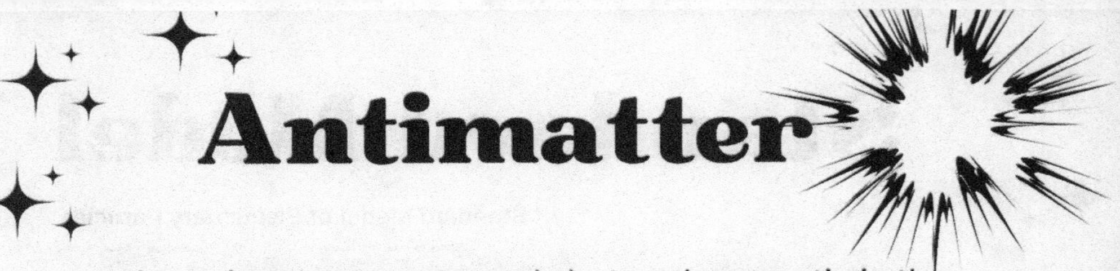

The Universe was extremely hot and energetic in the moments after the Big Bang. Particles of matter and their corresponding antiparticles were constantly being created in equal parts and annihilating. Antiparticles or antimatter are the opposite of the matter we know that makes up everything we see in the Universe.

Matter	**Antimatter**
Proton	**Antiproton**
Neutron	**Antineutron**
Electron	**Positron**

No joke, these are the names of the antiparticles. The atoms that make up matter and antimatter have the same mass but have opposite charges, which is why they annihilate upon contact and convert to energy as gamma-ray photons. Their existence may sound farfetched since we seem to be a Universe filled with matter, but antimatter is as much of a property of the quantum laws that govern this Universe as matter is. Antimatter matters.

But if they were made in equal parts, why are we a Universe filled with matter? It's still a cosmic mystery, but as the Universe began to cool and expand further, they were no longer being made in equal parts. For every 2 billion antiparticles, there were 2 billion *and 1* particles! This is known as baryogenesis. It paved the way for particles to collide to create all the objects we see in space.

By the way, this all happened fast as hell. From their creation out of energy, constant annihilation, and baryogenesis, it was only a fraction of a second after the Big Bang!

Whatever it was that caused this asymmetry did us a favor. We wouldn't exist without it.

Antimatter can be created through high-energy processes like particle collisions in particle accelerators and cosmic rays that come through our atmosphere. They are currently used in PET scans and could be a future rocket propulsion source. If we ever want to use it on a large scale, storing it properly without letting it come into contact with matter would need to be figured out.

Side note: antiparticles are a part of the standard model.

Superposition

It sounds sexual, but it's not. You may have heard of the famous Schrödinger's cat experiment. Is the cat alive or dead? Don't worry; no cats were harmed, as it was a thought experiment to explain the idea of superposition in quantum mechanics. It's also often used to reference anything that's a paradox.

Erwin Schrödinger was a Nobel Prize-winning physicist who worked alongside Einstein and other notable scientists during the era of quantum mechanics discoveries. Although he may be most famous for his cat experiment in pop culture, he won the Nobel Prize in 1933 for his contributions and discoveries in finding that particles like electrons behave as particles and waves. Remember, he created the wave equation, which calculates a particle's state based on probability.

Anyway, his cat experiment came in 1935 to highlight quantum mechanics' strange and counterintuitive aspects when applied to our everyday macro world. Basically, he was making fun of how crazy QM is. It involves putting a cat in a sealed box (cats love boxes) with a radioactive atom and a vial of poison. According to quantum mechanics, until an observation is made, the cat is in a state of <u>superposition</u> of being both alive and dead simultaneously due to the uncertain state of the radioactive atom. It's not until you open the box that superposition collapses into an outcome - the cat is alive or dead.

Superposition is where a quantum system can exist in multiple states at once. When an observation or measurement of that system is made, it "collapses" into a single state.

For example, suppose you have a car that's black AND white. When you open the garage door and see the car, it's now black OR white. But when you close the garage door and don't see it, it's black AND white as it's in a state of superposition. The whole system is in multiple states until an observation is made. This tends to have spiritual implications for people, but I'll leave that up to you.

Superposition is a cornerstone of quantum mechanics and emphasizes how crazy the quantum world truly is!

Double-Slit Experiment

The famous double-slit experiment broke scientists' minds. It showed that some particles behave like particles AND waves. This is known as particle-wave duality. Why is this such a strange thing? Well, we don't see this in the macro world. A ball behaves like a ball. When you throw it, it doesn't also act like a wave. A particle should be at a particular point in space and time and not spread out like a wave.

When you drop two rocks near each other in a lake, their ripples overlap. In physics, this creates what is known as an interference pattern.

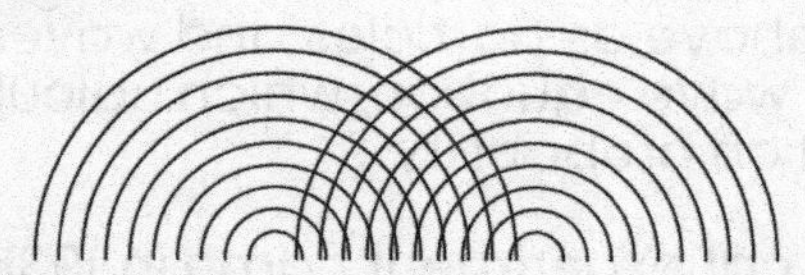

SET-UP: a barrier has two narrow slits and behind it is a screen that can detect when particles hit it.

- You shoot particles, like electrons one at a time from a particle source toward the barrier with the two slits.
- If you think of electrons as classical particles like tiny balls, you'd expect them to pass through one of the slits and hit the screen as two separate lines of impacts behind the slits. You'd see this if you threw balls through two openings in a wall.
- However, when you perform the experiment, you see an interference pattern generated by waves emerge on the screen as more and more electrons hit it. Wtf? Instead of two separate lines, you see a pattern of alternating bright and dark bands.
- Each particle was in a state of superposition, and their wave functions interfered with each other to create the pattern.
- It gets even weirder when you close off one of the slits because the interference pattern gets made again! It's as if each electron interferes with itself.

This experiment clearly highlighted particle-wave duality as these electrons behaved as waves... when they weren't being watched.

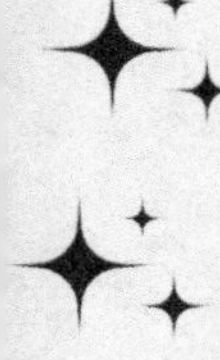

To See or Not to See...

The double-slit experiment keeps getting more bizarre.

When the particles weren't watched, observed, or measured, they exhibited wave-like patterns, as I just described.

But when a detector was set up to watch the particles, they changed their fucking behavior!

They almost conformed to how things work in our classical world and behaved as particles. Instead of an interference pattern, they formed just two separate patterns on the back screen behind each slit.
Again, wtf?

To reiterate:

When not observed

When observed

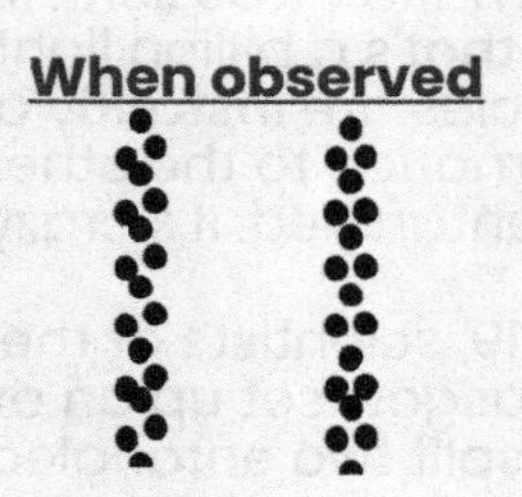

When observed, the particles behaved just as we'd expect them to. The mere act of observation collapses the wave function, forcing the particle to "choose" its definite state, left slit or right slit.

With no observation, they remain in a state of superposition as they go through both slits simultaneously. Their wave functions overlap and interfere constructively and destructively, creating the interference pattern.

Why? No one knows! It's absolutely bizarre. If you ever figure it out, a Nobel Prize awaits you.

Entanglement

Quantum entanglement is an experimentally proven phenomenon in quantum physics. I hate using this word when discussing science, but it seems like magic. Actually, much of what happens in the quantum world seems like magic since particles behave in a way that we're not accustomed to witnessing in the macro scales.

<u>Entanglement</u> is where two or more particles are interconnected no matter the distance between them. When one of the particles does something, it seems to instantaneously affect the other, even if they're billions of light years apart. It's like how twins *claim* they can sense when the other is distressed.

Does entanglement violate the cosmic light speed limit if they're instantaneously affecting each other? No. This part might hurt your head (more). Don't worry, it hurts mine too. Despite the instant effects of entanglement, the properties of relativity are not violated because, technically, no information was sent. Imagine one particle on each end of a pole that's a billion light years long. The pole spins and both particles are instantly affected. One particle didn't "send" information to the other and violate the speed limit. They're just entangled. It's crazy stuff! And... we even have a picture.

In 2019, scientists at the University of Glasgow set up an experiment that split two entangled photons. One was sent through a four-phase filter while the other didn't. Yet they both went through the same phase changes. Imagine if you and a friend got separated in a haunted house, but you both screamed, cried, and peed at the exact same time.

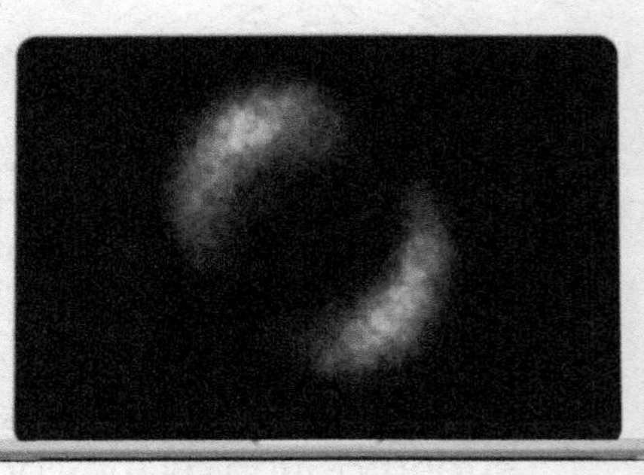

The experiment actually captured four images of four different phase transitions. It's quite remarkable! (It's either that or someone kissed a photocopier.) The more we understand the quantum world, the better our technology will get. Computers will be on a whole new level of processing information. I'm not a computer geek, so I won't delve into that area, but I know they will make our current computers seem archaic.

Regardless, there's still much to learn about a phenomenon that even Einstein called "spooky action at a distance."

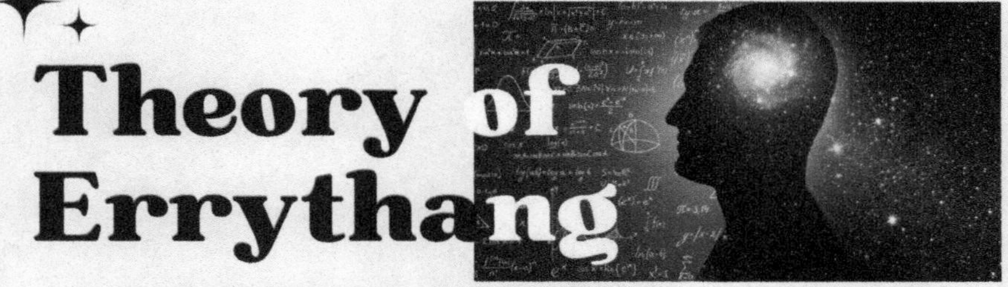

Theory of Errythang

"To read the mind of God."

In 1979, three physicists won the Nobel Prize for combining the weak and electromagnetic forces into the *electroweak force*. There are efforts to combine that with the strong force for an *electronuclear force*, but it has yet to happen. The final piece would then be to combine that with gravity into one unifying theory of everything that could explain the function of the entire Universe.

Relativity and quantum mechanics are incredibly successful in their ways but are incompatible in other ways. Nature, at its core, is simple. This is one Universe. We shouldn't need different theories to explain it. We're missing something here.

So far, it's been damn difficult to combine the macro (gravity) with the micro. That's why some are looking into quantum gravity. The search is on for a gravity-force-carrying particle.

- As I mentioned, a boson is a force-carrying particle. Things don't just happen in the Universe. A force doesn't just happen. Subatomic particles make things happen. Bosons are the particles that make the weak, electromagnetic, and strong forces happen. *Boson* also sounds like an insult, which I love.

Finding the hypothetical graviton, a boson for gravity would solve this. Scientists predict their existence only because bosons make the other forces happen, so why not gravity as well? If they find it, this macro force can finally fit into the micro world, thus unifying the four fundamental forces.

This doesn't mean that we could now rule the Universe. There could be advancements in technology and such in time, but ultimately, we'd be able to simplify this seemingly complex Universe down to one rule.

There's a reason why the phrase "theory of everything" is synonymous with Steven Hawking. He tried to find one grand unifying equation but could not, which shows its difficulty. But hopefully, it will happen in my lifetime. That would be one of the greatest moments in scientific history! And yes, another Nobel Prize would be in store.

Debunking

Flat Earth

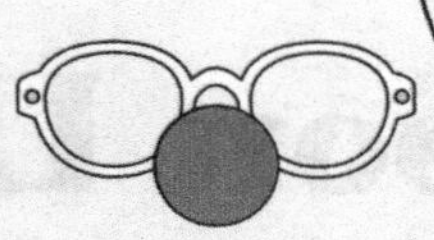

Here we go. Can I just say that I hate that I have to do this? Unfortunately, in 20 fucking 23, it's necessary, and my social media audience tends to love it.
Proof that the damn Earth is round.

- **Gravity**: wants to pull everything to a center. In a spherical Earth, that would be the center of the sphere. In a flat Earth, it would still be a center, but towards the center of the dish-shape. This doesn't make any fucking sense because if you lived in a country on the outer edges of this dish, you'd be pulled in sideways. I'm pretty sure apples don't fall sideways in Australia, but flefers don't believe Australia (or gravity) is real anyway.

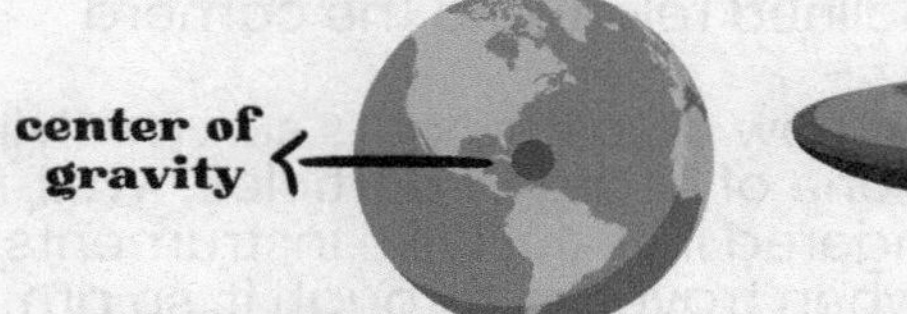

- **Lunar eclipses**: when the Earth's shadow falls onto the Moon, you see the mother fucking curve.
- **The horizon**: watch a ship with a sail set off towards the horizon, and you'll see the mast and the sail disappear last. If the Earth were flat, the whole thing would disappear beyond viewing simultaneously.
- **Shadows**: Greek mathematician Eratosthenes knew in 240 BCE that he could measure the angle and length of shadows cast by the Sun at the same time in different locations, and they would differ. That's because the Sun would hit both places at different angles due to the curvature. If it were flat, the length of the shadows would all be the same.
- **ISS**: at this point, we have images of the Earth, but flefers don't give a shit and say they're fake anyway.
- **Sunsets**: watch the same sunset from two different elevations. Start with the lower, then run to the higher elevation, and you will see the Sun reappear. It wouldn't do this if the Earth were flat.
- **Line of sight**: concerning the above point, you don't even need a sunset. If you were standing on the ground in a wide field, your line of sight would extend if you climbed a tree as more of the landscape concealed by the curvature would appear. Try that on a flat Earth; nothing would change, you idiots.
- **Privilege**: the fact that people are able to sit there and debate this crap shows just how privileged their lives are. Go focus on real problems in the world.

Moon Landing

Flat Earth is dumb as FUCK, but calling the Moon landing a hoax is insulting.

- **CoNvErGiNg LiNeS**: deniers say that the fact that some of the shadows on the Moon weren't parallel was proof of more than one light source. As in, it was filmed in a studio. WRONG. Even on Earth, the shadows created by the Sun aren't always parallel. Converging shadows are based on your perspective, and how the surface is inclined relative to the camera and light source.
- **VaN aLlEn RaDiAtIoN bElT**: how do astronauts survive flying through this dangerous zone of charged particles? Well, it's not like the astronauts lingered in it. Just like instruments on satellites are protected when traveling through it, so are the astronauts, who fly through it very quickly. It's the same way you can quickly walk across a fire pit.
- **No StArS iN iMaGeS**: basic ass photography would tell you about aperture, the space through which light passes in the camera. The Moon reflects a lot of light, so if you wanted to capture more surface features, you'd need a tiny aperture. Letting in this small amount of light wouldn't allow the stars to show up, obviously. So stupid. (I'm getting so irritated).
- **FlAg WaViNg In ThE wInD**: this has to be the dumbest one. They think they're being clever when they "call out" the waving flag as there's no wind on the Moon. Do these assholes actually watch the footage before opening their yaps? When Buzz Aldrin was SCREWING the flag in, NO SHIT it's going to wave back and forth! That's basic laws of motion. Once he walks away, the flag isn't fucking waving anymore ughhh.
- **We NeVeR wEnT bAcK**: Apollo 17 in 1972 was the last mission to land people on the Moon, and we didn't go back because NASA's budget got cut and priorities changed. We are going back this decade with the Artemis Mission.
- **RUSSIA FUCKING ADMITTED IT!!**
- At the time of writing this book, India successfully landed a spacecraft on the Moon's south pole. Idiots couldn't tell that the "footage" of the landing was obviously an *intentional* animation. We don't have cameras planted there waiting for it to land. SMDH.

Can we please stop shitting on the amazing achievements of these brave and intelligent people?

CGI

#CGI

The amount of times I've read that in my comments when presenting new space images is crazy. Even though these people mean it insultingly, they're not entirely wrong.

No shit computers generate these images. Astronomy is all about incorporating computer technology. What, you want people to hand draw a galaxy? Then they'd get hate for it being an "artist conception." So you're damned either way.

Anyway, this is how <u>image processing</u> works. For instance, Hubble looks in the visible, ultraviolet, and infrared parts of the EM spectrum. It already sees in true color (what it would look like to you if you had gigantic pupils many inches across), but filters are added to allow wavelengths that are beyond visible light to pass through.

Broadband filters let in a lot of light, while narrowband filters let in specific wavelengths corresponding to elements like hydrogen, oxygen, sulfur, etc.

Hubble takes multiple photos of the same object through these different filters. Scientists combine these images to make one comprehensive color image. They can also brighten or dim certain parts.

"False color" images are the ones that include wavelengths from beyond what we can see. Since we obviously can't see them, colors from the visible spectrum are added so we can.

The result? A shit ton of data is collected, allowing us to see things beyond our genetic limitations. The ultimate goal is to learn as much about an object as possible. It's not about you! So again, if you think you're throwing shade by saying these images are "CGI," you couldn't be more blind.

Cat's Eye Nebula was assigned red, blue, and green colors to the filters to distinguish between 3 narrow wavelengths of red light that correspond to hydrogen atoms, oxygen atoms, and nitrogen ions. These wavelengths are barely distinguishable to humans. Without assigning these colors, we wouldn't be able to see the differences between these molecules in the nebula.

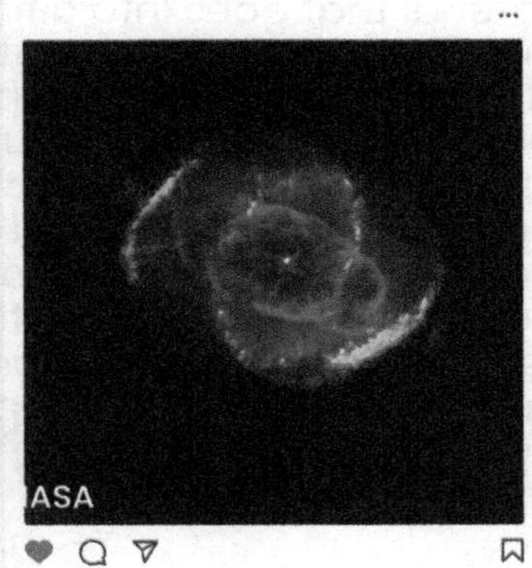

Defunding

One of the most discouraging things people say is that we spend too much money on space endeavors. The irony is that they often say this on social media with profile pictures that are selfies (you'll see why). So here are the many ways funding space institutions has benefited humanity. Oh, and let me just say that NASA's budget is ≤ 1% of the entire fucking federal budget!

- In the 70s, NASA's budget was cut down since we got to the Moon and the mission was accomplished. Therefore, the agency turned its efforts to Earth. We learned a lot about our climate and atmosphere.
- Speaking of atmosphere, we learned a lot about how climate change works by studying Venus, which basically has climate change on crack.
- Obviously, we know what's happening in the atmosphere at any given moment thanks to satellites. This helps us predict natural disasters and provide advanced warnings.
- We also have GPS because of satellites. Furthermore, they use general relativity (see General Relativity page).
- Studying quantum mechanics leads to quantum computers so conspiracy nuts can bitch and complain much more quickly.
- In the medical field, technology developed by the space program has led to CAT scans, artificial limbs, digital imaging breast biopsy systems, and tools for cataract surgery, to name a few.
- Other inventions include: dust buster, camera phones (this answers that ironic statement above), scratch-resistant lenses, athletic shoes, land mine removal, foil blankets, water-purification system, jaws of life, ear thermometers, home insulation, wireless headset, memory foam, freeze-dried food (not Tang, that's a myth), baby formula, computer mouse, and portable computers.
- The space program also developed the flame-resistant material that goes into a firefighter's suit.
- One day, humans will inevitably leave this planet. It's just within our nature. Mars is the next likely stop. Technology has allowed us to create fucking oxygen on Mars.
- Pondering the Universe makes you a better person.
- Space is collaborative and unifying.
- Space sciences provides jobs.
- You really want to defund the scientific institutions directly responsible for how comfortable your life has become? Instead of allocating their puny funds elsewhere, ask yourself, wtf are YOU doing to make the world a better place?

Yeah. So stfu.

Final Thoughts

This book is my best effort to share a subject I love, hoping you'll love it too. Some of it might have gone over your head, and that's okay. I tried to simplify the topics, but we are still dealing with some of the Universe's most complex, challenging, and mind-blowing concepts. Let them spark your interest, and trust that you will eventually grasp them with enough research and repetition!

This book is also a shoutout to those who dare to see the world for its possibilities through science, especially in this unfortunate and absurd age of antiscience. People often don't seem to understand why we spend money on space agencies because they don't think the science directly affects their lives. Despite the plethora of ways that I explained that it does, humans are naturally curious creatures. We can't help it, nor should we. We'll forever be on a quest to understand our journeys through the cosmos and where they all began.

I've always felt that pondering the Universe is as much of a philosophy as a science. (Did I say this already?) Nothing else can make you feel so small and insignificant yet simultaneously invoke such awe, wonder, and hope. Again, don't let the difficulties of the subjects deter you. None of these subjects were new to me, but even I learned and relearned so much while writing this book. I believe that thinking about things greater than ourselves will make this world a better place, and what better place to start than by being *spaced out*?

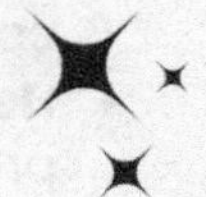

Quick Facts

- **Absolute Zero:** lowest temperature possible with no motion or heat. 0 Kelvin, (-273.15°C, -459.67°F).
- **Actual density:** the amount of mass in the Universe spread out over its volume.
- **Age of the Universe:** ~ 13.7 billion years old.
- **Aperture:** diameter of the light collecting region, the lens or mirror.
- **ASI:** Italian Space Agency.
- **Astronomy:** the branch of science that deals with celestial objects, space, and the physical Universe as a whole.
- **AU:** average Earth to Sun distance, 1 astronomical unit ~ 93 m miles (149 m km).
- **Boson**: force and energy-carrying particles.
- **Celestial Sphere:** an abstract sphere where the sky can be perceived as being projected upon, with the Earth at the center.
- **Chromatic aberration:** the fringing of colors as light from different wavelengths gets split and arrives at different angles.
- **Classical mechanics:** the branch of physics that deals with the motion of objects under the influence of forces.
- **CNSA:** China National Space Administration.
- **Constant:** unchanging physical quantity believed to be universal.
- **Cosmological constant:** denoted by Λ, it represents the mysterious force of dark energy and the evolution of the cosmos.
- **Critical density:** the amount of matter needed to stop the expansion of the Universe.
- **Ecliptic:** orbital path of the planets around the Sun, and the apparent path of the Sun across the sky over a year.
- **Ellipse:** in astronomy is the geometric shape the planets make as they revolve around the Sun, which is at one focus.
- **Entropy:** degree of disorder or uncertainty.
- **ESA:** European Space Agency, consisting of 22 countries.
- **Escape velocity:** the speed an object must obtain to "break free" of another object's gravitational attraction.
- **Event horizon**: point of no return around a black hole.
- **Galaxy:** collection of millions - trillions of stars, planets, gas, dust, and dark matter. Most house a supermassive black hole.
- **Gravitational waves:** ripples in the fabric of spacetime caused by cataclysmic astronomical events.
- **Heisenberg Uncertainty Principle:** there is an inherent limit to how precisely certain pairs of properties, like position and momentum, can be simultaneously known for particles.
- **Hidden sector:** a hypothetical collection of quantum fields and their corresponding particles that have yet to be observed.

Quick Facts

- **Hubble constant:** expansion rate of the Universe.
- **ISRO:** Indian Space Research Organization.
- **JAXA:** Japan Aerospace Exploration Agency.
- **"k":** 1,000.
- **L2, Lagrangian point:** where gravitational forces and the orbital motion of a body balance each other. Allows spacecraft to hover.
- **Light:** encompasses the entire electromagnetic spectrum.
- **Light speed:** 186,000 miles (300k km) per second. Or about 7.5 times around the Earth in one second.
- **Light year:** distance light travels in a year ~ 5.8 t miles (9.3 t km).
- **Megaparsec:** 3.26 million light years.
- **Moon:** a natural satellite that orbits a planet.
- **NASA:** National Aeronautic and Space Administration.
- **Nebula:** a giant cloud of gas and dust.
- **Nuclear fission:** splitting the nucleus of an atom into two or more smaller nuclei, releasing a lot of energy in the process.
- **Observable Universe:** ~ 93 billion light years in diameter.
- **Parallax:** the apparent shift in position of a nearby object against a distant background.
- **Parsec:** 3.26 light years.
- **Planet:** large, round bodies that revolve around a star, and are the most gravitationally dominant object in their line of orbit.
- **Photon:** a massless particle of light (a type of boson).
- **Precession:** slight change in orbit or rotation of a body over time.
- **Quantum mechanics:** the study of the microscopic world, atoms, molecules, subatomic particles, etc.
- **Redshift:** the wavelength of the light gets stretched towards red as the object moves away from you.
- **Relativistic:** speeds that are more than a significant fraction of the speed of light. This is when relativity theory functions apply.
- **ROSCOSMOS:** Russian Federal Space Agency.
- **Singularity:** a point of infinite density and gravity.
- **Spectroscopy:** the study of the absorption and emission of radiation by matter.
- **Standard candles:** objects with known luminosities.
- **Star:** a large and naturally self-luminous body that has acquired enough mass to ignite nuclear fusion.
- **Theory:** observation of the natural world that is repeatedly tested and corroborated and hasn't been able to be disproven.
- **Time dilation:** time slows down the faster you go.
- **Zodiac:** a belt across the sky that extends 8 degrees above and below the ecliptic that contains 13 constellations.

Made in the USA
Monee, IL
19 November 2024